Lionel Swift

The Manual of the Hydrometer

Lionel Swift

The Manual of the Hydrometer

ISBN/EAN: 9783744763912

Printed in Europe, USA, Canada, Australia, Japan

Cover: Foto ©berggeist007 / pixelio.de

More available books at **www.hansebooks.com**

THE MANUAL

OF THE

HYDROMETER;

CONTAINING

ITS HISTORY; PHILOSOPHY, MODE OF GRADUATING SCALE,
APPLICATION TO TECHNICAL AND GENERAL PURPOSES;
WITH RULES, WORKED EXAMPLES, AND
COMPLETE TABLES.

CHAPTERS ON THE

EFFECTS OF SURFACE CONDENSERS;

THE CAUSE OF

OXYDATION AND DEPOSITIONS

IN MARINE BOILERS,

ITS PREVENTION AND CURE;

PRIMING; MANAGEMENT OF BOILERS

AND SUPERHEATERS.

By LIONEL SWIFT,

INSPECTOR OF MACHINERY AFLOAT, ROYAL NAVY.

SECOND EDITION.

REVISED AND ENLARGED.

PORTSMOUTH :

GRIFFIN & Co., 2, THE HARD.

[PUBLISHERS TO H. R. H. THE DUKE OF EDINBURGH.]

LONDON :

SIMPKIN, MARSHALL & Co.;
E. & F. N. SPON, CHARING CROSS; IMRAY & SON, MINORIES;
PHILIP & SON, LIVERPOOL.

1870.

PREFACE.

SOME apology is due to our readers for the delay that has occurred in the appearance of this second edition so long after the announcement of its being in preparation. To those familiar with the duties of fitting out a large and somewhat new type of ship, most probably the explanation will suggest itself: the nature of these public duties allow but narrow intervals unclogged with professional anxieties and special inconveniences: and the faculty of self-abstraction and command of thought is not easy to untrained and new authors. However, there being a continued demand for the old work, which had for some time been out of print, I was induced to prepare a second edition. Feeling that I had something to say about the cognate matters referred to in the supplementary chapters of this volume, and having a desire to curtail, re-arrange, and otherwise alter and amend the old and former part, I willingly undertook the responsibility of appearing again before the public, on whose kind forbearance I rely, that the same indulgence will be extended both to the views put forth, and tne obvious deficiencies of authorship, which is so generously and usually accorded to merely practical, though earnest workers.

It is some source of congratulation that nearly all the suggestions prominently made in the first edition, although then but in very limited use and adoption, are now become matters of established and orthodox practice.

I would particularly mention the high importance and economy of surface condensers, the value of "*scum collectors*," and the views, I believe for the first time then propounded, of the *modus* of calcareous depositions in Marine Boilers. These two last features become of the highest interest in face of the practical inconveniences found to occur, and more particularly referred to, in the additional chapters of this edition.

PRINTED AT THE OFFICE OF THE PUBLISHERS.

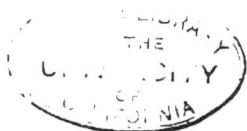

INTRODUCTION.

EXPERIMENTAL science has demonstrated that in very many instances of chemical combinations there is an attendant change of a physical character totally unaccountable, and frequently of a nature apparently paradoxical. Thus, some combinations exhibit a considerable decrease in bulk, whilst others increase in volume by mixing. There are other extraordinary changes resulting from some combinations, such as the alterations of temperature, the relative hardness, the color, the freezing point, and other changes dependent on the ratio of the peculiar ingredients; but it is more particularly the change of the resultant volume in combinations, that we propose remarking upon. Dr. Ure gives the following list of ingredients in alloys of metal that respectively increase or decrease in volume by mixing, viz. :—

That INCREASE in Volume.	That DECREASE in Volume.
Gold and Silver	Gold and Zinc
,, ,, Lead	,, ,, Tin
,, ,, Iron	,, ,, Bismuth
,, ,, Copper	,, ,, Antimony
,, ,, Iridium	,, ,, Cobalt
Silver ,, Copper	Silver ,, Zinc
Iron ,, Bismuth	,, ,, Tin
,, ,, Antimony	,, ,, Bismuth
,, ,, Lead	,, ,, Antimony
Tin ,, Lead	Copper ,, Zinc
,, ,, Palladium	,, ,, Tin
Zinc ,, Antimony	,, ,, Palladium

And several other combinations.

But perhaps the most extraordinary instance of change in volume occurs in mixing iron with platina.

B

If ten cubic inches of iron be mixed with one and a quarter cubic inches of platina, the bulk of the compound is only nine and three quarter cubic inches, or $10 + 1\cdot25 = 9\cdot75$, exhibiting a concentration in volume equal to more than thirteen per cent. Again, in the mixing of two parts of brass and one part of tin, whose respective specific gravities are $8\cdot006$ and $7\cdot363$, the sp. gr. of the mixture becomes $8\cdot917$, whereas, if each had retained its former bulk, the sp. gr. would have been $7\cdot7916$. A mixture of equal parts of the above ingredients should have the sp. gr. $7\cdot684$, but it is really $8\cdot441$. It is also remarkable in this instance, that mixing brass with this lighter metal has made a composition harder than is due to the relative hardness of the metals, and *more* dense and heavier than the original brass could be made by any ordinary hammering or compression.

A further instance is afforded in mixing $16\frac{1}{2}$ ounces of alcohol with 20 ounces of water, when the condensation is equal to about $\frac{1}{16}$ of the whole bulk of the ingredients. So that a pint of proof spirit would not realise a quart of mixture.

So also, 100 ounces of water, mixed with 34 ounces of common salt; the condensation in volume is equal to four per cent. of the sum of their respective volumes.

And in the case of turbith-mineral, if some is added to a narrow-necked bottle filled with water, the water, instead of rising in the neck of the bottle, sinks considerably, and the two ingredients occupy less space than the water did alone.

Enough has been shown in these instances to satisfy us that the usual mode of determining the relative amount of ingredients cannot be deduced from the specific gravities of the constituents in the compound, and the specific gravity of the compound itself; and that Archimedes must rather have overstepped the practical fact in his assumption that he had discovered the means of detecting the extent of dishonesty practised in the manufacture of the famous crown of King Hiero. The desirability of a mathematical solution of these physical laws suggest an ample field for investigation and research. Some combinations have been made the subject of most elaborate and complete experiment, more especially, perhaps, the experiment with varied mixtures of alcohol and water, prosecuted for the guidance and information of the Board of Excise; but hitherto these, and indeed all experiments on this subject, have only resulted in the application of particular empirical formulæ, adapted each to its respective case. In applying ourselves to the consideration and bearing of sea-water brines, to the practice of Naval Engineering, our object will be to examine first, the different forms of Hydrometers, their principles as applied to determine specific gravities, and the mode and philosophy of their graduations, more especially for testing the brines of sea-water; and secondly, to examine the change in volume in brines, under different proportions in the amount of contained salt, and with variations of temperature; and further to deduce some practical formulæ applicable to the working of marine boilers.

"HYDROMETERS."

THE HYDROMETER was known and used by the ancients, having been originally invented by Hypatia, the accomplished daughter of Theon, a mathematician of Alexandria.

The form of Hydrometer first used was one with an uniformly graduated stem, a pear-shaped form of ball, and a weighted ball underneath.

The Hydrometer is based on the principle, that the weight of a floating body is equal to the weight of the quantity of liquid which it displaces. Therefore, the weights required to sink an Hydrometer equally far in different liquids, will be directly as the densities of the liquids; and the Hydrometer will sink in different liquids in an inverse proportion to the density of the liquids. Hence these two facts have given rise to different kinds of Hydrometers :—*The first*, with a graduated scale on the stem, and the volume of immersion varying; *the second*, with a fixed point of floatation, and brought to that point by adjusting weights to the stem; and the *third*, a combination of weights and graduated scale on stem.

For anything like an extended range of specific gravities, and great delicacy in observing, the last-mentioned one alone is to be depended upon.

We do not, however, intend to refer to all the various Hydrometers that special requirements may have brought into use, but only such as may be of practical service to the Engineer, and have met with general approval and adoption. First of which is—

THE COMMON HYDROMETER.

Common Hydrometers have only a graduated scale marked on the stem, and vary in form and material, as thought most desirable by the makers. In most cases they are not arranged for the indication of pure specific gravity as such, but for the examination of particular liquors.

They generally consist of a ball, B (as in fig. 1), a slender graduated stem, S, and made steady by a weighted ball, A. If it is so loaded as to sink to the mark *o* on the scale, in the lightest liquid proposed to be measured by it, in a fluid of intermediate specific gravity, it will sink to some mark between *o* and the extreme of its range.

In this form of Hydrometer the weight is always the same, and the immediate information given by the instrument is that of different bulks with equal weights. Because the instrument sinks till the bulk of the displaced fluid equals in weight the instrument, and as the additions to the displaced fluids are all made by the stem, it is evident that equal distances on the uniform stem indicate equal additions of volume. Thus the stem becomes a scale of bulks to the same weight.

Now if such a stem be divided into equal parts, it will indicate the bulks in equi-different progression, but it will not indicate specific gravities in equi-different progression, because the specific gravity equals the weight divided by the bulk, and as the weight is always constant for the same instrument, its being divided by bulks decreasing in equi-different progression, the quotients, or specific gravities, will be what is called in harmonic progression, their differences continually diminishing.

The following example will illustrate our meaning, and afford the means of ascertaining the pure specific gravity from an instrument marked in equal divisions.

EXAMPLE.—Let the volume of a Hydrometer, up to the mark in which it floats in the liquid of the least specific gravity, = 5 cubic inches, the sectional area = .05 square inches, or $\frac{5}{100}$ of a square inch,

Then we see that the instrument may be regarded as an elongated tube of this .05 sectional area, and 100 inches in length (as 100 × .05 = 5 cubic inches). Now supposing it divided into tenths of an inch, then each *ten* of these divisions, or 1 inch in length, would equal .05, or $\frac{5}{100}$ cubic inches; and so on for any length of stem, as all specific gravities are inversely as their bulk. For let the weight of the instrument equal a, and let it float in fresh water up to the mark o.

Then $\frac{a}{5} = 1.000$, the specific gravity of water. And at another volume $\frac{a}{5 - .05}$ = specific gravity of the liquid. Then $\frac{a}{5} : 1.000 :: \frac{a}{5 - .05}$ = to specific gravity required. Thus $\frac{5}{5 - .05} = 1.010101 =$ the specific gravity for 10 divisions on stem. Similarly, $\frac{5}{5 - .1}$. $= 1.020408 =$ the specific gravity for 20 divisions.

As we stated, these specific gravities are in what is called harmonic progression; and it may be shown that to indicate specific gravities in equi-different progression, the divisions on the scale must be in harmonic progression, and with an increasing series of specific gravities, the spaces or divisions will be a decreasing harmonic series.

As for instance, given a Hydrometer to divide the scale, so as to indicate specific gravities from fresh water, or 1.000 to a specific gravity of 1.050, in equi-different progression, *i.e.*, as 1.000, 1.010, 1.020, 1.030, 1.040, and 1.050, the size and weight same as last example. Now as the sectional area of stem was .05 square inches, and the cubic contents of whole Hydrometer 5 cubic inches, we showed that it would equal a tube of the same sectional area as stem, and 1000 *tenths* of an inch long; thus, $\frac{1000 \times .05}{10} = 5$ cubic inches as the specific gravities of bodies when of equal weights are inversely as their bulk. For let a = the weight of the instrument, as before, and it displaces 1000 tenths of an inch when immersed in fresh water, (whose specific gravity is 1.000). Then as

$$\frac{a}{\text{whole volume}} : \frac{\text{specific gravity of water}}{1.000} :: \frac{a}{\text{smaller volume}} : \left\{ \begin{array}{l} \text{specific gravity of the} \\ \text{liquid.} \end{array} \right\}$$

Now, in the case of this example, we want to find the respective volume corresponding to 1.010, 1.020, and so on; then we shall have $\dfrac{a}{1000}$: 1.000 :: $\dfrac{a}{1000 - x}$: 1.010 ; where x = the difference of volume corresponding to 1.010 of specific gravity ; then to find x we have, $\dfrac{1.010a}{1000} = \dfrac{1.000a}{1000 - x}$ dividing by a, and clearing fractions, $1000 = 1.010 \times (1000 - x)$ $\therefore$ $x = 1000 - \dfrac{1000}{1.010}$, = 9.9 tenths of an inch the first division ; then length of second division by same rule will be $1000 - \dfrac{1000}{1.020} - \left(1000 - \dfrac{1000}{1.010}\right) = 9.607$ tenths of an inch. To further simplify the operation, let us calculate the respective distances from the mark o, the point to which the instrument sinks in fresh water :—

FIG. 2.

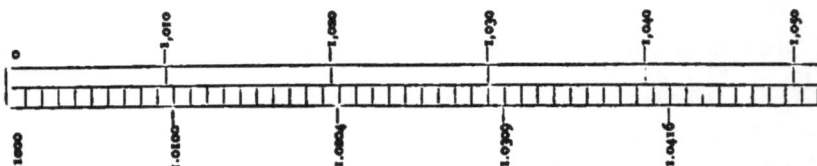

Then—

$1000 - \dfrac{1000}{1.010} = 9.9$ tenths of an inch from o = 1.010

$1000 - \dfrac{1000}{1.020} = 19.6078$,, ,, = 1.020

$1000 - \dfrac{1000}{1.030} = 29.1$,, ,, = 1.030

$1000 - \dfrac{1000}{1.040} = 38.46$,, ,, = 1.040

$1000 - \dfrac{1000}{1.050} = 47.6$,, ,, = 1.050

and the respective distances between each gradation would be 9.901, 9.707, 9.519, 9.335, and 9.156 tenths of an inch.

We may show as follows, that no difference in the result ensues from regarding the instrument as an elongated tube, taking the actual contents as 5 cubic inches, and the sectional area as .05 square inches. Let x = the length in inches betwixt o, and the mark to indicate 1.040 specific gravity ; then $x \times .05$ = volume betwixt o and the mark required. Then $5 - .05x$ equal volume displaced of that specific gravity. Then as $5 \times 1.000 = 5 - .05x \times 1.040$

∴ x = 3.846 inches as before. If it was desirable to extend the range of an instrument such as we first regarded, that is 5 inches long in stem, same sectional area as before, and same entire cubical capacity, it would indicate at the extreme of its range as follows : 5 × .05 = .25 cubic inches capacity of stem betwixt o and the end of 5 inches length on stem, then 5 × 1.000 = (5 — .25. × the specific gravity of that bulk of an equal weight of liquid, or $\frac{5}{4.75}$ = 1.05263 : then to further extend the range, weights must be added sufficient to sink the instrument to the mark o, in a liquid of the specific gravity corresponding to the last indication without weights, or 1.0526 ; regarding the weight of the Hydrometer as 1. Then, let the additional weight = $\frac{1}{r}$. We shall then have $\frac{1 + \frac{1}{r}}{5} = \frac{1}{4.75}$ ∴ $r =$ 19. or $\frac{1}{19}$ of weight must be added (because weight divided by volume is equal to the specific gravity in all cases). Whence .05263r = 1, therefore $\frac{1}{19}$ of the whole weight of the instrument must be added to bring it down to o, when placed in liquid of 1.0526 specific gravity. And, with this additional weight, it will indicate at the extreme of its range (of 5 inches on stem) which will reduce the volume of fluid displaced to 4.75 cubic inches, thus $\frac{1 + \frac{1}{19}}{5}$: 1.0526 :: $\frac{1 + \frac{1}{19}}{4.75}$: specific gravity, when floating at the mark 5 inches from o. Whence $\frac{95 \times 1.0526}{90.25}$ = 1.108, the specific gravity of the fluid, when floating with the additional weights to the mark 5 inches from o. The specific gravity, corresponding to any other intermediate mark, may readily be calculated as in above and previous example. An instrument of this description, with the weight or weights inserted well down inside the stem, would form a very commodious Hydrometer for general use, combining, as it would, sensibility with a considerable range. The scale in Fig. 2, divided into equal parts, with their corresponding specific gravities on one side, and divided harmonically with the specific gravities in equi-different progression on the other side, will assist in making our meaning plain as to the method of graduating these instruments.

Although such rules as we have just laid down are the rules by which all scales must be divided, there are certain floating points which will be required (generally in practice) to be determined by experiment; and it will be advisable to take these points as remote from each other as possible; that is, in liquids of previously ascertained specific gravity near the limits of the range for which the instrument is intended.

When these two points are marked, and their distance apart measured, we shall be enabled to get the volume or capacity of stem, and this, divided by the measured length, will give the sectional area of stem in very correct terms, whence any other mark may be calculated as required.   The mode of proceeding to obtain the volume of stem contained between any ascertained limits of specific gravity would obviously be as follows, taking the same general data as before. Supposing we have a Hydrometer weighing 1266 grains, and the distance on stem between fresh water ·mark and the floating mark of a liquid whose specific gravity is 1.05263, is found to be 5 inches long;

gr. in 1 oz.   cu. in.   gr.   cu. in.

then as, 437.5 : 1728 :: 1266 : 5, the volume to the mark o, at which it floated in fresh water; then as 1.000 : $\dfrac{1266}{5}$ :: 1.05263 : $\dfrac{1266}{\text{volume at 5 in.}}$.   Whence, $\dfrac{5}{1.05263}$ = 4.75 the volume to mark 5;

then, $\dfrac{5 - 4.75}{5 \text{ in. long}}$ = .05 square inches, the sectional area of stem, as required, whence any other points may be calculated with the greatest exactness.

(2.)

NAVAL HYDROMETER.

The Hydrometer at present in use by Engineers in the Royal Navy (as originally constructed for and intended), is a most anomalous instrument for their requirements.   It was, and is, a *brewer's* instrument for ascertaining the strength of *worts*.   Its shape is similar to Figure 1, only the scale being divided harmonically.   It is arranged as follows : when it stands at 10, it indicates that a barrel of 36 gallons of the liquor weighs just 10lbs. heavier than a barrel of fresh water; and when the instrument floats at 20, then a barrel of the liquor would weigh 20lbs. heavier than the same bulk of pure water, and so on, each smaller division of the scale indicating an accession of weight of 1lb. in the 36 gallons; and from the coincidence of 10lbs., or 160 ounces in 36 gallons, being equal to the difference in weight of 36 gallons of sea-water, of average specific gravity, and 36 gallons *of fresh water* (as *we see from* the following proportion), thus: as $(277.274 \times 36) : 1000 \times 160 :: 1728 : 1027.7$ weight of a cubic foot of sea-water.) We say, from this coincidence are we indebted for the infliction of this arbitrary instrument on the Service for testing generally the brines of sea-water. But the parallel of its correct application ceases at the above point of sea-water. We might very naturally be disposed to imagine from the arrangement of the scale of this instrument, that, if its standing at 10 showed that the water contained $\frac{1}{a}$ of salt, then would $\frac{2}{a}$ of salt correspond to 20 on scale, and so on for the higher points, as 30, 40, &c.; and, indeed, we know it has been pretty generally received as such; but this assumption · involves very considerable error, as for instance, by reference to the Table A of specific gravities, densities, &c., at the end, we shall see that even at 20 we have an excess of nearly 4 ozs. of salt in a cubic foot, more than that mode of reading the indications would lead us to suspect; and at 30 the excess would be nearly 9 ozs. in a cubic foot.

This, in the more saturated brines, would be an increasing error, and let it be observed, on the unsafe side, in working boilers; for when we might apprehend an approximation to a dangerous state of saturation, we should really be far into it, and beyond all safe working limits.

We are usually disposed to regard leniently errors on the safe side of practice; but anything that would tend to lull our vigilance, and induce us to feel confidence when the crisis of prudent management is overstepped, cannot be too strongly reprobated, or too quickly remedied. And it appears to us that the requirements of an Engineer's Hydrometer are to indicate the proportion of solid matter liable to be deposited on the boiler, or the pure specific gravity, as such only, leaving the Engineer's knowledge, to supply the rest. We are, however, informed that another Hydrometer is about being submitted for use in H.M. Service, and we may express the hope that it will be so arranged as to meet the requirements of Engineers, and fulfil the conditions that philosophical considerations impose on such an instrument. We are sanguine in such a case, that the instrument will prove, in the hands of the educated and reflective body of Engineers, of valuable assistance in eliminating those mathematical conditions, connecting the varied circumstances of dilation, temperature, and capacity for heat, with their relative specific gravities, which at present appear to be so unsatisfactorily understood and regarded.

There are very many other Hydrometers, or, as they are usually called Salinometers, devised and much used with Marine Boilers.

The late Mr. Seaward's plan may be remembered; it consisted of a glass cylinder, something like a large guage glass, connected to the boiler, and inside this cylinder were one or two glass balls of a specific gravity, that, when they floated, should indicate when to blow off.

Mr. How's Patent Salinometer had also a considerable use both in the Royal Navy and otherwise. It consisted of an open vessel permanently connected by pipes with the different boilers.

The vessel is supplied with a Hydrometer graduated for 200° Fah., and is also accompanied with a Thermometer, to regulate the incoming brine to this temperature; and as the water is being continually supplied from the boiler, an overflow pipe is provided to the bilge.

"Gamble's Improved Patent Salinometer," and "Saunders' Salinometer" are more recent inventions, the latter an exceedingly concise and practical instrument, and as they are self-contained, they are preferable in this respect to "How's," avoiding the contingency of scalding from overflow of hot brine. The advantages in these instruments, are the facility of reading the indication at a glance, without the trouble of drawing and cooling down the brine. Some patentees, however, frequently claim considerably more, indeed almost a *specific* for preserving boilers, preventing scaling, and saving fuel; and sometimes they volunteer a little information more popular than correct.

Moreover, in addition to partaking of the errors mentioned as belonging to the common Naval Hydrometer, it has the further disadvantage inseparably connected with any Hydrometer for sea brines, that is graduated for anything but the standard temperature of 60° Fah. ; *viz.* the different amount of dilation in volume of brines normally differing in density, when subject to equal additions of temperature. Thus, brine containing $\frac{3}{a}$ of salt is affected to at least the amount of 2° on the Hydrometer betwixt the limits of 60° and 200° temperature more than brine containing $\frac{1}{a}$ of salt would be affected between the same limits of temperature. Therefore, if the instrument is arranged for sea-water, and stands at 3.0° of density at a temperature of 200°, we shall have an excess of 15½ ounces of salt per cubic foot more than indicated by the instrument, viz., from condensation in volume, 7 ounces per cubic foot ; and from difference in dilation, with 140° of temperature, 8½ ounces. This very loose system of notation leads to errors too great to be tolerated, and we had possibly better rely on the more primitive method of being guided by the boiling temperature to indicate the proportion of contained salt, than such lame indications as these instruments furnish. But, at any rate, we learn the fact, that *no* instrument will serve *both* for *varied* temperatures and *varied* densities ; and therefore we can *only rely* on an instrument graduated for the standard temperature of 60° Fah., and accompanied by tables for correcting the density as the observed temperature may differ from that normal point (for which see Tables.)

There is a further objection to these inaccessible instruments, that every Engineer's experience will have furnished many examples of— viz., the errors accruing from viscidity of the fluid, preventing the correct indication of any Hydrometer, and the frequent formation of air bubbles on the instrument, leading, to sometimes, great errors ; and both these sources of error can only be thoroughly obviated by an experienced observer shaking or working the instrument up and down until it settles at its proper level.

(3)

OTHER HYDROMETERS.

FIG. 3.

"*Nicholson.*"—This instrument is arranged to determine the specific gravity of solids, as well as liquids. Fig. 1 B. is a hollow ball, to which is attached a stem, s, supporting a dish, c, for receiving weights. Proceeding from the under side of the ball is the stem, carrying a heavy dish, D, preserving the stability of the instrument when it floats, and for holding any solid body, whose specific gravity is to be determined.

The instrument is floated in pure water ; and a weight of a 1000 grains is put into the dish c. Now the Hydrometer is so adjusted that it sinks to about the middle of the stem, and a mark, S, is made at this point.

To determine the Specific Gravity of a Liquid.—Place the instrument in the liquid, and place weights in the dish, c, until the mark S sinks to the surface of the liquid. These weights, added to the weight of the instrument, will be equal to the weight of the liquid displaced. But the weight of the Hydrometer, added to 1000 grains, is equal to the weight of an equal bulk of water ; therefore, the former sum divided by the latter will give the specific gravity of the liquid.

EXAMPLE I.—Let the weight of the Hydrometer equal 800 grains, the weight put on the dish equal 1090 grains (and, sinking to S), required the specific gravity of the liquid.

Here, $880 + 1090 =$ weight of displaced liquid.

„ $1000 + 800 = $ „ „ water.

∴ Specific gravity of liquid $= \dfrac{1890}{1800} = 1.050$ *Ans.*

EXAMPLE II.—Find specific gravity of liquid with above Hydrometer when 784 grains are put on c.

Here, $\dfrac{800 + 784}{1000 + 800} = .800$—Specific gravity required.

To determine the Specific Gravity of a Solid.—Place the Hydrometer in water, and put the solid in the upper dish c. Add weights to the dish until the mark *S* comes to the surface of the water ; then 1000 grains minus these weights must be the weight of the body. Let the solid now be put into the lower dish, D, and as before add weights to bring the mark *S* to the level of the water. Then 1000 grains minus these weights, equals the weight of the body in water ; and the weight of the body in air (by first proceeding), minus the weight of the body in water, equals the weight of an equal bulk of water ; and, as previously determined, the weight of the body in air divided by the weight of an equal bulk of water, is the specific gravity of the body. Thus,

EXAMPLE III.—A body was placed in the upper dish, and required 300 grains to sink the Hydrometer to its proper level ; and, when placed in lower dish, 400 grains were required to bring it to the same level. Required specific gravity of body.

Here, weight of body in air = 1000—300 = 700 grs.
Ditto ditto in water = 1000—400 = 600 grs.

Then, as before, $\dfrac{700}{700-600} = 7.000 =$ specific gravity of body.

DIFFERENT HYDROMETERS.

The divisions on the stem or scale are but rarely adapted to indicate specific gravity as such, but more usually adapted to the questions that arise in the peculiar business or purpose for which they are intended : thus, dealers in spirits use an instrument to indicate the amount of what is technically called "proof spirits," and for this purpose many instruments, very ingenious in their principle, have been brought into requisition. The one at present in use, and by which the amount of spirit-duty to be collected is estimated by the Excise Department, is known as "Syke's Hydrometer." It is arranged for ten weights, and its scale is divided into ten equal parts. The whole is adjusted for a temperature of 60° Fah., and tables are computed whereby the necessary corrections may be determined for temperature, above, or below that point ; the difference of this instrument and the one we suggested for marine practice, consists in the weights being placed outside of lower stem in Syke's, instead of inside tubes, as we propose. We rather consider this must be found inconvenient, frequently, as the sensibility must be impaired from the viscidity of the fluid acting on such irregular forms, as the additional weights would involve, and which would be avoided by placing the weights inside the stem, made hollow for the purpose, and secured with a cap to keep out water, or liquid.

There are, of course, many other Hydrometers, with special adaptations, that hardly fall within the scope of our subject to describe, on a purely Engineer's view of the matter.

Condensation of Volume.—In reference to the condensation of volume that takes place in one or both the primary constituents of brines, that is the pure water, and the salt held in solution. The amount of condensation appears the greatest in the very diluted brines; as we mentioned, that when a very small quantity of salt is added to a quantity of water in a bottle with a narrow neck, the compound sinks in the neck of the bottle, instead of rising, which of course is the result of a certain condensation taking place, evidently in this case in the water, and probably in both; but supposing for a moment that no change of volume took place in the water, then would the salt, in the case just mentioned, occupy less than no space, which of course is an impossibility.

We can ascertain by analysis the exact amount of each of these primary constituents, and of course the respective volumes of each, when existing separate; but we have no means of ascertaining the respective spaces occupied by each when compounded; but in endeavouring to illustrate the amount of condensation, we will again, for a moment, regard it as if the whole condensation took place in the salt; and, viewing it in this manner, it will be apparent that the condensation appears greatest in proportion to the extent of dilution; and gradually diminishes as more salt is added. Perhaps it may not be worth consideration, when this change takes place, gradually or abruptly, as in setting of mortar, and other instances. From our previous observations it will be seen how very rapidly this condensation is affected in the earlier stages of dilution; and the sulphates of soda, and many others, present curious modification of similar results.

The specific gravity of the whole of the saline portion of sea-water, in a perfectly dry state, and slightly pressed together sufficiently to form a compact mass, we found in varied trials not to exceed 2.000 ; and when saturated to its greatest extent, it appears to rise to 2.850 ; and when saturated to a state corresponding to four times the density of sea-water, the specific gravity becomes 3.160 ; and when containing twice the amount of salt that sea-water contains, the specific gravity of the salt is increased to 3.476, and the specific gravity of the salt when existing in the proportions found in sea-water is further increased to 3.933 ; and again, when further diluted to a specific gravity of compound of 1.0042, the specific gravity of the salt becomes about 11.000, and still regarding it with the same view of change, the specific gravity of course would thus increase to an unassignable value, when diluted to its greatest extent.

As we before stated, the physical law regulating this change in volume of the compound is not yet known. It has been suggested that the curve of a hyperbola, whose asymptotes are an angle of about 30°, would afford a very near approximation to the change of relation betwixt the weight of a constant bulk of the compound and the weight of the dry saline constituents contained therein ; that is, if the abscissa were made to represent the weight of contained salt, then the ordinates would be equal to the weight of a constant bulk of the compound ; but there is considerable practical difficulty in applying this mode of interpolation, arising from the great disparity necessarily in the lengths of the respective ordinates and abscissa. An empirical formula might, however, readily be constructed with an exponent that should be some function of the weight of a constant bulk of the compound, and coincide in result with the weight of its contained salt as derived from observation, sufficient for the usual limits of marine practice. The following simpler formula will, however, be found very near the observed result betwixt the limits of 1.027 and 1.100 of specific gravity. Thus (observed specific gravity) — 1.027 × 1.5 + 36.06 = weight of contained salt in cubic foot of compound.

Influence of Temperature on Volume and Specific Gravity of Brines.—
The great dilation of water, when subjected to heat, has been long
noticed and duly recorded, as also of brines fully saturated with salts;
from which it appears that, while fresh water expands 0.04332 or
$\frac{1}{28.08}$ of its volume, brine, when fully saturated, and of a specific gravity of 1.215, expands 0.05198, or $\frac{1}{19.23}$ of its volume, by being
heated from a temperature of 39° to 212° Fah.; and further, fresh
water is but slightly affected in volume by slight additions of heat to
the temperature of its maximum density. Brines are far more sensible
of the additions of heat at similar low temperatures, and comparatively
are not so much affected at high temperatures as fresh water. Thus,
between the temperature of 40° and 60° fresh water expands .00083
of its volume, and betwixt 192° and 212° it expands .00747 of its
volume.

The dilation in this case is nine times greater for the 20° of temperature nearest its boiling point than it is for the 20° nearest its
maximum density; whilst a brine fully saturated would expand betwixt the temperature of 40° and 60°, and the temperatures of 192°
and 212°, only in the proportion of about 1 to 1.34; while betwixt
the temperatures of 110° and 130°, the proportion or ratio of expansion is about uniform for both liquids. Thus it will be seen, that not
only the ultimate amount of expansion of a volume of these brines,
but the *rate* of expansion at or near the limits of the above range in
temperature, is entirely dependent on its intermediate state of saturation; and we again see how impossible it is for any Hydrometer to
answer for brines in different states of saturation, *at varying temperatures*, unless corrected by tables, the result of observed experiment.
And it will also be clearly perceived that no Hydrometer can be correct for different densities, if graduated for a temperature beyond 60°
Fah. The practical effect of temperature on brines, in relation to the
amount of salt contained at a given specific gravity, is greater than
what is due to condensation of volume, at about the average state of
saturation at which boilers are usually worked, as we showed, in reference to water at 3.0, by Hydrometer giving an error in the proportion of 8 ½ to 7 ounces of salt per cubic foot under the circumstances

mentioned. Engineers have always been in the habit of making a certain allowance for the influence of temperature on the density of brines when observing, some allowing 10° and others 12° of temperature to 1° of the Hydrometer; as, for instance, if the brine was at a temperature of 80° above the temperature at which the instrument was graduated for, and the Hydrometer stood at 12° in such brine, they would, allowing 10° of temperature to 1° on the Hydrometer, then say the brine was of a density of 12° + 8°, or corresponding to 20° on the Hydrometer when at its proper temperature. However well this method may be adapted for ready practical purposes, it will be seen that it can only furnish an approximation, which will necessarily be impaired under almost the majority of circumstances. Hence, with a view of extending our practical acquaintance with this subject, we, as long ago as 1849, made a series of observations with a Hydrometer graduated to 55° of temperature, with brines of different stages of saturation betwixt the temperatures of 55° and 180°.

Our reason, then, for not continuing the observations to a higher temperature was the very erratic and uncertain indications at these high temperatures ; but we have recently gone over a somewhat similar series in another form, in which we think we have avoided the irregularities at high temperatures, to which our experiments with the Hydrometer were subjected. In these, our later experiments, we have made use of a small glass bottle containing about two cubic inches, in the neck of which we inserted, perfectly water-tight, a small glass tube four inches long.

The different brines to be tested were prepared from clear pellucid sea-water boiled down to the required density. In conducting the experiments every care was taken, by a very slow and gradual process in lowering the temperature, to secure as perfect an identity as attainable in the temperature of the brine in the bottle and the surrounding water in which the thermometer was inserted. It is only proper to observe that no consideration has been taken of the expansion of the bottle, the object being more to obtain the *comparative practical* dilations, at different normal specific gravities, and as such, we confidently submit the result, as seen in Table B, trusting it will be obvious and easy of reference.

THE AVERAGE SPECIFIC GRAVITY OF SEA-WATER.

In attempting to fix the average specific gravity of sea-water, we found on collating a variety of authorities, the discrepancies existing most discouraging; but, if the relative amount of contained salt for varying weights of a constant bulk of sea-water is correctly determined, the mean specific gravity of the ocean resolves itself into little more than a question of scientific satisfaction.

However, in fixing the mean and normal density of sea-water at 1.0277, we have been guided considerably by personal observation in various parts of the world, but more implicitly have we relied on the extended and accurate experiments of the late Dr. Marcet on this subject, which are thus stated:

" 1st.—That the Southern Ocean contains more salt (for equal bulks) than the Northern Ocean in the ratio of 1.02919 to 1.02757.

"2nd.—That the mean specific gravity of sea-water near the equator is 1.0277, or intermediate between the northern and southern hemispheres.

" 3rd.—That there is no notable difference in the character of the constituents of sea-water, under different meridians.

" 4th.—That there is no satisfactory evidence that sea-water under great depths is salter than at the surface.

" 5th.—That the sea in general contains more salt where it is deepest and most remote from land, and that its saltness is always diminished in the vicinity of large masses of ice.

" 6th.—That small inland seas, though communicating with the ocean, are much less salt than the ocean.

" 7th.—That the Mediterranean contains rather a larger proportion of salt than the ocean."

Further, Dr. Marcet gives the following as the specific gravities of the various seas, as ascertained at a temperature of 60° Fahrenheit.

NAME OF SEA	SP.GR.	NAME OF SEA.	SP. GR.
Arctic Ocean	1.02664	Sea of Marmora	1.01915
Northern Hemisphere ...	1.02829	Black Sea	1.01418
Equator	1.02777	White Sea	1.01901
Southern Hemisphere ...	1.02882	Baltic	1.01523
Yellow Sea	1.02291	Ice Sea-Water	1.00057
Mediterranean	1.02930	Dead Sea	1.21100

Mr. Mallett, in his more recent report for the "British Association," makes the specific gravity of sea-water 1.0278, and Admiral Fitzroy, the Director of the Meteorological Department of the Admiralty and Board of Trade, in his report, just published, states the specific gravity as 1.027; but for a variety of reasons, in addition to those assigned, we are disposed to adopt 1.0277 as the most correct exponent of the specific gravity of sea-water generally, and in referring to sea-water we must be understood as meaning the above density at a temperature of 60° Fahrenheit.

AMOUNT AND CHARACTER OF SALINE INGREDIENTS.

Notwithstanding the very able and decisive analysis of several very eminent chemists, the proportion of saline matter in sea-water is still given in many popular works on "Steam" in the most loose and uncertain terms, and too frequently erroneous in amount.

We are usually met with the assertion that sea-water contains $\frac{1}{32}$ or $\frac{1}{33}$ of salt, and in some very recent works it is almost amusing,—the very evident uncertainty of the authors as to whether it is $\frac{1}{32}$ of its weight or of its volume, but a compromise seems to be effected by stating it at its weight in one place, and its volume in another.

We have been at some pains to trace to its paternity this common but fallacious $\frac{1}{32}$ propensity, and we find in Vol. 111., page 1432, of Dr. Thompson's *System of Chemistry*, that Lord Mulgrave found the water at the back of Yarmouth Sands to contain $\frac{1}{32}$ parts by weight of saline ingredients. This authority, confirmed as it was by Dr. Murray's experiments of the water in the Frith of Forth, in the earlier days of engineering obtained the highly respectable sponsorship of such men as Watt, Tredgold, and others, and it is doubtless owing to the high position of this society that its legitimacy as an engineering fact has passed current so long without challenge.

The correct proportion in different localities has been determined by Thomson, Watson, Marcet, Schweitzer, Urd, Laurens, and more recently by Mr. Mallett, and we cannot see why the correct notation should not assume its proper place instead of the fallacious $\frac{1}{32}$ proportion, which can only lead the student at the outset of his investigations in this subject into error and confusion.

The tables of relative proportion of contained salt for different specific gravities, &c., (see Table A,) are based more particularly on the experiments of Drs. Robinson and Watson, and agree remarkably with Dr. Marcet's observations, and (we may perhaps be pardoned for modestly adding) are confirmed, as far as our limited observations have extended. Dr. Ure gives the following proportions of saline ingredients from different localities in 1000 parts by weight. "The largest proportion of salt held in solution in the open sea is 38, and the smallest 32. The Red Sea contains 43 ; the Mediterranean 38 ; the British Channel 35.5 ; the Arctic Ocean 28.5 ; the Black Sea about 21 ; and the Baltic only 6.6."

We may observe that the Baltic is considerably affected in its proportion of salt held in solution by the direction of the prevailing wind, and the same cause would no doubt affect in a smaller degree the proportion of contained salt in the British Channel. The specific gravity of the Baltic water as affected by the wind is exhibited as follows :—

(KIRWAN's "GEOGRAPHICAL ESSAYS," PAGE 356.

Sp. Gr.	Direction of Wind	Sp. Gr.	Direction of Wind
1.0039	Wind at East	1.0118	Storm at West
1.0067	,, ,, West	1.0098	Wind at N.W.

Dr. Schweitzer's analysis of the water at Brighton gives $\frac{1}{28.36}$ of saline ingredients. Dr. Ure and Mr. Mallett make the proportions in the British Channel $\frac{1}{28.17}$ of salt matter. Mr. Laurens makes the Mediterranean to contain $\frac{1}{24.4}$ of solid matter. Mr. Mallett states it to amount to $\frac{1}{25.38}$ of saline matter (all by weight).

However, leaving these little discrepancies, we shall submit the extended analysis by Dr. Schweitzer of the water at Brighton, and by Mr. Laurens of the water of the Mediterranean, as types of the character of the saline constituents of sea-water generally.

ENGLISH CHANNEL AT BRIGHTON, BY DR. SCHWEITZER.

1000 parts by Weight contained as follows.		
Water	964.74372	
Chloride of Sodium	27.05948	
„ „ Magnesium	3.66658	
„ „ Potassium	0.76552	Specific Gravity 1.0274
Bromide of Magnesium	0.02929	
Sulphate of Magnesia	2.29578	
„ „ Lime	1.40662	
Carbonate of Lime	0.03301	
Total weight	1000.00000	

The water examined was stated to be quite pellucid, with but the slightest trace of organic matter.

1000 parts by Weight contained as follows.		
Water	959.06	
Chloride of Sodium	27.22	
„ „ Magnesium	6.14	
Sulphate of Magnesia	7.02	
„ „ Lime	0.15	Specific Gravity 1.0293
Carbonate of Lime	0.09	
„ „ Magnesia	0.11	
Carbonic Acid	0.20	
Potash	0.01	
Total weight	1000.00	

Dr. Schweitzer remarks in reference to these analyses that it will be observed that the Channel water contains six times as much lime in solution as the water of the Mediterranean, which he attributed to the local geological features of the Channel, composed as it is of a bed of lime.

ON THE DEPOSITS IN MARINE BOILERS.

All sea-water holds therefore (with many other salts) a proportion of lime in solution ; and if this caustic earth were not disposed of, the effect of evaporation would, in time, unfit the sea for the purposes or as a medium of life, and the Foraminifera, are most active and widely diffused instruments by which this soluble caustic lime is precipitated in the condition of a mild, insoluble carbonate of lime.

It is only when brines become highly saturated, that any deposition of chloride of sodium (or common salt) takes place ; but other deposits take place considerably before this period, as, for instance, according to Professor Farraday's experiments, sulphate of lime is deposited when the proportion of saline ingredients to the water is as 1 to 10, or at a specific gravity of 1.075, or 2.7 by Naval hydrometer ;

and Dr. Davy states that this sulphate of lime "constitutes without any exception" the nature of the deposits examined by him ; so much so, that unless chemically viewed, the other ingredients may be held to be of little moment. However decisive the above high authorities may be regarded, the experience of engineers furnishes innumerable instances of very considerable and constant deposition in marine boilers, when they had every assurance that the saturation of the boilers had never attained the above density; and indeed in boilers evolving steam, the tendency to deposition may be observed in the slight white film with which the flues, &c., are coated after but a few hours steaming, and when the boilers have never attained a much higher density than that due to ordinary sea-water. This may probably arise from the carbonic acid gas, which holds the carbonates of lime and magnesia in solution in the form of bicarbonates, becoming evolved with the steam, (as from the repulsive action of heat on this gas, it would naturally separate,) and thus these carbonates of lime, &c. are deposited, however low the state of saturation may be. This early deposit is no doubt slight at the commencement, but there is little question that it would materially accelerate subsequent forma-tions by acting as a nucleus with other salts.

This view of the modus of lime deposits is very strongly corrobo-rated by the condition in which we find our domestic kettle. And we have known cases in which the anxiety of the engineer to prevent scale formations has wrongly led him to sustain the very low density of 15 by Hydrometer, when steaming at an average speed on a voyage, and which has resulted, as might have been expected, in disappointment at finding more than an ordinary coating of lime on the tubes and furnaces, &c. This can hardly be wondered at when we consider that 50 per cent more feed was used to keep the point of saturation down to 15, than would have been required to keep the boilers at a fair and safe point, say of 20, and that nearly all the lime entering with this feed is most probably precipitated almost immediately on mixing, with the mass of water in the boiler. The object of preventing these deposits, has called into requisition no end of patented processes, many of which, if not postively dangerous in their nature, are at least very equivocal in their results.

" Riley's Composition " has found an extensive use in the Navy, and been most favorably reported on.

Feeding the boilers from the bilges has also been strongly recommended by some engineers of great practical experience; and we have certainly seen boilers that have been continually at work for more than twelve months, remarkably free from incrustation, or deposit of any kind, which the Engineer attributed entirely to the constant use of bilge water as feed, when lying with fires banked. However, be that as it may, it is perhaps worth the consideration how far the decomposed, fatty, and organic matter, usually common to bilge water, would furnish the requisite chemical re-agent with lime and magnesia in an incipient state of deposit, and thus arrest its formation on the flues, &c.

Baker's " Patent Ant Incrustator " is very extensively used ; we must admit however, we have met with very conflicting testimony as to its value, and this can hardly be wondered at when we remember how different it is always to comply with the often highly subtle conditions required to bring into efficient action magnetic influence, as is sought by this patent.

We have seen an account of some wonderful results, both in cleaning a boiler badly incrustated, and in preventing its formation ; by the use of a bar magnet suspended inside the boiler, between the surface of the water, and the top of the boiler, the south pole of the bar being connected with the shell of the boiler, while the north pole is supported by an insulated hook, or other arrangement—the effect of the magnet so fixed is remarkable.

The incrustation or scale on the inside of the boiler falls of, and no more is deposited, and this cleansing effect is maintained so long as the bar remains in proper magnetic condition.

In very large boilers, two or more bars may be required, placed side by side, to produce the proper effect, or the bar must be made an electro-magnet, in which case any amount of cleansing or preventing power may be maintained.

It is quite obvious, that only this early tendency to deposition has to be combatted ; the " *blow off* " will most effectually do the rest.

As further bearing on the prevention of these calcareous formations, we take the liberty of making an extract from a note of Mr. Armstrong's in Tredgold's (new edition), where he says, speaking of some experiments at Mr. Scott's works on the River Wear :—

" The result of these experiments was the adoption of the plan of placing one or more small collecting vessels within each boiler, and then, instead of blowing the sedimentary deposit out from the general mass of water in the boiler, the blow-off pipe was attached to the inner vessel, which stood at some distance from the boiler bottom. This blow-off being discharged several times a day, was found to be quite effectual in freeing the water in the boilers from nearly all its previous impurities.   The unlooked for, and at that time singular fact, first noticed by Dr. Clanny during these experiments, and which it is the purpose of this note to mention, as it has not yet been followed up in all its important consequences to Steam Navigation, is, that in using salt-water in steam-boilers, the brine drawn from the inner collecting vessels is always in a much more concentrated state than the surrounding mass of water.   So much is this the case, in fact, that I have frequently found it to be nearly a fully-saturated solution, which chrystalized immediately after exposure to the atmosphere, while the remainder of the water in the boiler was in its ordinary working state."

To find the Amount of Brine to be blown out.

RULE.—Multiply the evaporation in any given time, by the density of the feed water, and divide the product by the difference in density of the feed and the density to be sustained in boiler.

NOTE.—Let d = density of feed (by Table A. Column 4).

,, d' =   ,,   brine and water in boiler.

In any ⎧ ,, f = amount of feed in cubic feet.

given ⎨ ,, b =   ,,   brine   ,;   extracted.

time. ⎩ ,, e =   ,,   evaporation in cubic feet.

,, n = ounces of salt in a cubic foot of sea-water, density 1.000.

EXAMPLE I.—Supposing the evaporation of a set of boilers is 210 cubic feet per hour, and it is required to sustain a density of 1.500 by feed water of the same density as sea-water, viz., 1.000, required the amount of brine necessary to be abstracted per hour?

Here, by rule, $\dfrac{\text{evaporation} \times \text{density of feed}}{\text{density of boiler}-\text{density of feed}} = \dfrac{210 \times 1.000}{1.500-1.000}$

= 420 cubic feet of brine to be blown out per hour; and 420 + 210 = 630 cubic feet = feed per hour.

EXAMPLE II.—Required the amount to be blown off, when the density to be sustained in boiler is 2.500, the evaporation and density of feed as last example?

Here $\dfrac{210 \times 1000}{2.500-1.000}$ = 140 cubic feet per hour.

EXAMPLE III.—Given the evaporation 500 cubic feet per hour, find the amount in weight to be blown off, when the specific gravities of boiler and feeds are respectively 1.060 and 1.020?

Note continued.

Let s = specific gravity, corresponding to d density.
  ,, s' =   ,,   ,,   ,, d' ,,

Then will f, d, n = weight of salt going into boiler with feed in time given; and b, d', n, = weight of salt extracted with brine to sustain uniform density in boiler. These must evidently be equal, f, d, n, = b, d', n.

Hence $b = \dfrac{fd}{d'}$........(equation 1).

But 1000 $f s$ = total weight of feed.

And 1000 $b s'$ =   ,,   ,,   brine blown off.

Then 1000 fs — 1000 $b s'$ = weight of water evaporated.

Hence as $\underset{\text{Wt. of cubic ft.}{\ \ \text{of water.}}}{1000}$ $\underset{\text{Cub.ft.}}{1}$ :: $\underset{\text{Wt. of evaporation}}{1000\,(fs - b s')}$: $\underset{\text{In cub. ft.}}{e}$ ∴ $b = \dfrac{fs-e}{s'}$............

(equation 2).

But, by equation 1 & 2 $\dfrac{fd}{d'} = \dfrac{fs-e}{s'}$ whence $f = \dfrac{e d'}{d' s-d s'}$, and substituting this value of f in equation 1, $b = \dfrac{ed}{d'' s - d s'}$ which is the correct expression for the rule; but the omission of any consideration of s & s' involves only an inconsiderable error in the result, and may conveniently be left out in practice when the rule becomes as given above. Although a nearly similar investigation to the above

Here densities, corresponding to above specific gravities, are 2.288 for 1.060 and .704 for 1.020. (See Table A.)

Then $\dfrac{500 \times .704}{2.288 - .704} \times \dfrac{1060}{16} = 15301$ lbs. per hour of brine to be blown off.

EXAMPLE IV.—Required the density of the boilers when $\frac{1}{3}$ of the feed is blown off, and density of feed is .704?

The evaporation in this case will be $\frac{2}{3}$, then by Rule $\dfrac{\frac{2}{3} \times .704}{\text{boiler density} - .704}$ $= \frac{1}{3}$ or $2 \times .704 + .704 = 2.112$ density of boiler.

EXAMPLE V.—In a set of boilers evaporating 395 cubic feet per hour, required the amount requisite to be blown off per hour to sustain a density of 2.200, with feed from hot well corresponding to a density of .850 ?—*Ans.* 249 cubic feet nearly.

EXAMPLE VI.—Find the density of the boilers when $\frac{1}{4}$ of whole feed is blown off, and density of feed = 1.000?—*Ans.* 4.000 density.

EXAMPLE VII.—Find the density in above example when $\frac{1}{2}$ of whole feed is blown off?—*Ans.* 2.000 density.

To find Boiling Temperature corresponding to any Density and Pressure.

The temperature at which water or brine boils in the open air, is dependent on the amount of barometric pressure, and the quantity of salt held in solution. The boiling temperature of fresh-water, when the barometer stands at 29.8 inches, is 212° Fah., and a difference of

---

Note continued.

appears in Messrs. Maine and Brown's very excellent work on the *Marine Steam Engine*, I had arrived at this solution long before it appeared in that work, and had indeed shown the investigation in 1849 to one, if not both, of the above authors, considerably prior to its first insertion in their second edition.

I should not, by any means, wish it to be inferred that the above authors adopted my notation, simple as it is, without acknowledgment; but I claim, at least, the originality of arriving at the same result by independent means, and considerably antecedent to their publication of the same.

E 2

one inch of mercury in the barometer affects the boiling temperature 1.76° Fah. at about the mean atmospheric pressure ; but this relation of pressure to boiling temperature by no means holds good under pressures, such as boilers are usually worked at, the augmentation in boiling temperature appearing continually to decrease in amount in a ratio inverse to the increase of pressure.   *We say somewhat inversely, for, be it observed, that here again,* the theoretic or physical law, connecting the pressure with its corresponding boiling point, is unknown ; we therefore rely for the connexion on the recorded results of the extended experiments on this subject.

The results obtained by the Academy of Sciences in Paris have mostly obtained the confidence of scientific men, and will be found in the list of Tables in Carr's *Synopsis of Practical Philosophy*, 24mo. For Specific Heat, *Rudimentary Treatise on Steam, &c.*, vols. 78-79, in Weale's Series.   Tredgold and Hodgkinson on the *Strength of Cast Iron and other Metals*, 8vo., 1860-61.   Various empirical formulæ have been constructed to approximate very closely to the obtained result of experiment.

We shall here only give the one proposed by Mr. Tredgold for this purpose, which is as follows :—

$$\left(\frac{103 + t}{201.18}\right)^6 = p \quad \begin{cases} \text{when } t = \text{temperature (sensible)} \\ \text{„ } p = \text{pressure in lbs. per sq. inch, therefore,} \end{cases}$$

201.18 $(p^{1/6})$—103 = t or temperature of water in boiler.

But, for the water containing the usual salts of sea-water, we find that the boiling temperature in the open air is raised 1° Fah. for each addition of 2.58 per cent of saline ingredients, which is equal to 1.4 very nearly for each degree of density ; then, for sea-water brines, the above formula would stand thus :—

$(P^1 \times 201.18)$—103 + 1.4 d = pressure in lbs. per sq. inch.

When d = density, or proportion of salts in solution (by Table A), in a boiler actually working with sea-water, the formula would then become, when t was the observed temperature $\left(\dfrac{103 + t - 1.4d}{201.18}\right)^6 =$ pressure in lbs. square inch.

The operation of resolving these quantities, involving as they do the 6th root (or square root of the cube root) of the pressure, is very considerably facilitated by the application of logarithms, as the following examples will illustrate.

EXAMPLE I.—Observing the temperature of a boiler to be 258.5° Fah., and the density 2.200, required the total pressure under which the boiler is working? Here by formula,

$$\left(\frac{\overline{103 + 258.5} - \overline{2.2 \times 1.4}}{201.18}\right)^6 = \text{pressure.}$$

Or $\left(\dfrac{358.42}{201.18}\right)^6 = \text{pressure}$, then by logarithms—

Log. ⁻358.42 = 2.554392

„ 201.18 = 2.303584

= .250808

6

Pressure = 31.97 = 1.504848 from which we see the pressure would be nearly 32 lbs. per square inch on the boiler.

EXAMPLE II.—Required the temperature of evaporation of the water in boilers working at a pressure of 30 lbs. per square inch, and a density corresponding to 2.500 (by Table A)?

Here log. of p., or 30 lbs.	= 1.477121
Which, divided by 6	= .246197
Adding log. of 201.18	= 2.303584
or 354.6	= 2.549771

Then, 354.6 − 103 + (2.5 × 1.4) = 255.1° temperature required.

THE CAPACITY FOR AND SPECIFIC HEAT OF BRINES.

The terms capacity for heat and specific heat, like specific gravity, have a relative signification, that is, without affirming the absolute amount of a body's specific heat, it is compared with the amount of heat contained in the substance referred to as unity. For ponderable bodies, water is usually adopted as the standard; but the specific heat of most bodies changes in itself by additions of temperature, and some considerably more than others: thus, fresh-water has a considerably less capacity for heat at about its boiling temperature, than at a lower temperature, such as 40° to 60° Fah.; but a brine, or other substance of a less nominal specific heat, is more constant in its specific heat. Dr. Dalton found that the heat required to raise water 5° in the lower part of the theometric range would raise it 6° in the middle of the same range.

There exists, no doubt, a considerable relation betwixt the specific heat of a body and its dilation in volume between the same limits of temperature; but it scarcely evinces that uniformity requisite for a definite law.

MM. Dulong and Petit's experiments established an indisputable relation between the specific heats, and the primitive, combining atoms of all simple bodies, which, however, with the present arbitrary atomic weight of binary compounds, is scarcely applicable to deducing the specific heat of these compounds by this process. We have, perhaps, appeared to use the terms capacity for heat and specific heat somewhat indiscriminately: the fact is, there is no very great confusion arising from regarding them as identical. The former more properly means the relative powers of bodies in receiving and retaining heat, in being raised to any given temperature. The latter term, *i.e*, specific heat, applies to the actual quantities of heat so received and retained in any given weight; so that to convert the specific heat of equal weights into the specific heats of equal volumes, it is only necessary to multiply the specific heat, or capacity for heat of equal weights, by the specific gravity of substance. The result will be the specific heat of equal volumes. Thus, the capacity for heat of a pound of mercury is .033, when the capacity for heat of a pound of water is 1.000; then, these multiplied by their respective specific gravities, will give the specific heat of equal volumes of the substances; therefore, the amount of heat it would take to raise a pound of water 1° of temperature, would raise 30 lbs. of mercury 1° of temperature, whose capacity for heat is represented by .033, and water represented by 1.000; but the amount of heat that would raise a cubic foot of water 1° of temperature, would only raise barely 2.25 cubic feet of mercury 1°. The respective capacities for heat of equal weights, or equal volumes, do not change the terms from capacity for heat to specific heat, as erroneously stated by some authors. The specific heat of sea-salt is stated as .20; fresh-water being 1.00: therefore, any amount of heat that would raise a pound weight of water through a certain range of temperature, would raise 5 lbs. of salt through the same range. Now, if we take 100 parts by weight of sea-water (specific gravity 1.0277), we find that it contains 96.39 oz. of water, and 3.61 oz. of salt.

Then, $96.39 \times 1.000 = 96.39 =$ Total heat in water,

and $3.61 \times .200 = \underline{\quad .72 \quad}$,,   ,,   salt.

Whence ...... 97.11 ,,   ,,   compound.

Then, $\dfrac{97.11}{96.39 \times 3.61} = .971$ specific heat of compound for equal weights.

The results obtained in this way coincide so remarkably with the recorded observations of Gadolin and others in the specific heats of brines of varying density, as to confirm the above inductive mode as the correct mathematical exponent of the relation between the specific heat and the specific gravity of brines. To change the specific heat of equal weights to the specific heat of equal volumes, we stated it was only requisite to multiply, by the respective specific gravity. Thus, .971 × 1.0277 = .998, the specific heat of equal volumes of sea-water, an equal volume of fresh-water still being 1.000. But it will be seen we shall obtain the same result, if we take the respective amounts of fresh-water and salt in a cubic foot multiplied by their respective specific heats, and divided by 1.000.

Thus sea-water contains 990.58 of water, and 37.72 of salt. Then $\dfrac{(990.58 \times 1.0) + (37.12 \times .20)}{1000} = .998$ specific heat of equal volumes ; and as we usually estimate feed, evaporations, &c., in terms of cubic feet, this will afford the most available means of considering the question of the relative economy of heat as applied in practice.

It may be interesting here to examine and compare the relative economy of using fresh-water and sea-water in a boiler. For this purpose we shall assume that the feed is of the density of sea-water, and in the proportion of water to salt, as above stated. Then we shall have in a cubic foot of sea-water, as 1.000 :· 1728 : : 990.58 : 1711.72 cubic inches of fresh water in a cubic foot of sea-water—then as 1711.72 : 1 : : 1728 : 1.0095, which will give the amount of sea-water to be supplied as feed to equal one cubic foot of fresh-water. Then we shall assume the heat required to convert these into steam, as being equal for equal volumes of fresh-water, that is raised from 212° to 213.4°; the heat for conversion to steam is less by the amount of this difference. Therefore 1.0095 × .998 = 1.0075 = the amount of heat to produce one cubic foot of water in the form of steam, against 1000, the amount to produce one cubic foot of water in the form of steam from fresh-water. Therefore the relative economy of producing steam from fresh-water and sea-water (not, of course, here considering the loss from blowing out), is as 1.000 to 1.0075, or only .75 per cent. of loss from using sea-water.

When two given substances are mixed together, the temperature of the compound is the sum of the weights multiplied by their respective temperatures and specific heats, and divided by the sum of the weights multiplied by their respective specific heats.

EXAMPLE I.—Let a pound of water at 50° temperature, be mixed with two pounds of mercury at 160° temperature. Required temperature of compound when the specific heats are respectively 1.000 and .033. Here $\dfrac{(1 \times 50 \times 1) + (2 \times 160 \times .033)}{(1 \times 1) + (2 \times .003)} = 56.8°$ the temperature required.

EXAMPLE II.—Supposing 10 lbs. of fresh-water at a temperature of 50° to be mixed with 10 lbs. of brine, containing 13.5 per cent. of salt, and at a temperature of 200°, required the temperature of compound. The specific heat of fresh-water is 1.000, and to find specific heat of brine, we have by previous rule, $\dfrac{(100 - 13.5) + (13.5 \times .20)}{100}$

= .892 specific heat of brine. Then $\dfrac{(10 \times 1.000 \times 50) + (10 \times .892 \times 200)}{(10 \times 1.000) + (10 \times .892)}$

= 120.7° temperature of compound.

EXAMPLE III.—Let the fresh-water in the above example have a temperature of 200°, and the brine 50°—other circumstances the same —to find temperature of compound. Then

$$\frac{(10 \times 1.000 \times 200) + (10 \times .892 \times 50)}{(10 \times 1.000) + (10 \times .892)} = 130° \text{ temperature of compound.}$$

TO FIND THE PROPORTION OF FUEL LOST BY "BLOWING OUT."

RULE.—Multiply the latent heat of the steam by the evaporation per hour, to which add the difference of temperature of water in boiler, and water of feed, multiplied by the amount of feed per hour, and the capacity for heat (of equal volumes) of feed water, and call the result A. Then take the same difference of temperature of boiler and feed, which multiply by the amount of brine blown out per hour, and by the capacity for heat (of equal volumes) of brine—call this B. Then as A : $\frac{8}{10}$ of coals used per hour : : B : to coals lost by blowing out.

> NOTE.—Let E = Evaporation per hour in cubic feet
> b = Brine blown out
> f = Feed = (E + b)
> T and t = Temperature of boilers and feed
> c and c' = Capacities for heat of feed and brine respectively in equal volumes
> L = Latent heat of the steam due to boiler pressure, which will be 1114—.695 (T—1.4 d,) when d = density of boilers
> F = Fuel used per hour
> Then (T—t) fc = Heat expended in raising feed water to temperature of boiler
> And L E = Heat expended in evaporating from temperature of boilers
> ∴ L E + fc (T—t) = whole heat utilized
> And b c' (T—t) = Heat discharged in brine
> Then as L E + fc (T—t) : F : : b c' (T—t) : fuel lost per hour in blowing out.

We must observe here, that this apparent loss of fuel is considerably in excess of the actual loss, as in the above we have taken no consideration of the heat discharged up the funnel, or lost by radiation.

The amount of loss from these sources, is estimated in the best arranged boilers, as quite equal to 20 per cent., or, 80 per cent. only of the fuel is utilized in the water.

Having regard, therefore, to these considerations, the more approximate loss becomes reduced to the following :—

As L E $+fc$ (T—t) : $\frac{8}{10}$ F : : $b c'$ (T—t) : to fuel lost, which is the expression for the above rule.

EXAMPLE.—Given a set of boilers, evaporating 250 cubic feet per hour, under a total pressure of 30 lbs. per square inch, the temperature of feed water being 100°, and its density .850, the density sustained in boiler being 2.000. Required the loss of fuel per hour from blowing out, when the total consumption per hour is 2000 lbs ?

To find amount of brine extracted per hour— $\dfrac{250 \times .850}{2000 - .850} = 185$ cubic feet per hour nearly. Then feed will amount to 250 + 185 = 435 cubic feet per hour. The temperature due to 30 lbs. pressure is 251.6, from which we find that temperature of boiler will be 251.6 + (2.000 x 1.4) = 254.4°, and to find latent heat of steam to accord with Regnault's late experiments, 1114—(.695 x 251.6°) = 939°.

Then by Rule—

$$939 \times 250 + (254.4 - 100) \times 435 \times .998 = 234750 + 67130 =$$
$$301880 = A, \text{ and } (254.4 - 100) \times 185 \times .993 = 28364 = B.$$

Then as 301880 : $\frac{8}{10}$ 2000 : : 28364 : 150 lbs. loss per hour. The per centage will be as $\frac{8}{10}$ of 2000 : 150 : : 100 : 9.37 per centage of loss in fuel from blowing out.

The above example is about the usual average of practice; we, therefore, see that the total per centage of loss from blowing out and using sea-water (see page 36), amount to about 10.12, a per centage of loss that very strongly suggests the economy that would result from surface condensers and fresh-water.

CAUSES AND PREVENTION OF BOILER PRIMING.

CAUSES AND PREVENTION OF BOILER PRIMING,

PRIMING is a very complicated phenomenon. Molecular force, the nature of which is so highly complex, undoubtedly plays an important part; and the fact of so little being understood of its action, might reasonably deter us from discussing those features of priming with which we are more conversant. Priming in excess is the bodily upheaving of large masses of the water, when its vis inertia is overcome by the momentum of the escaping steam; this is accelerated in the first instance by the steam bubbles rising through the surface, surcharged with watery particles combining with the surface water in violent ebulition; and thus joining in the direction acquired by the escaping mass, is therewith bodily carried up and along, constituting what is understood as PRIMING: the severity and excess of which is only limited by any mechanical retarding contrivance, or by limiting the amount of energy in the conditions causing the primary tendency. These influencing conditions may be found partly in the form of the boiler more particularly at its water surface, for if we have a considerable heating surface, and a contracted water surface, it is inevitable that the effort made by the rising bubbles of steam to concentrate themselves in a smaller area, will create a highly ebullitionary condition of surface; and probably for some distance we shall have a considerable mass of water in a condition of density but little heavier than surcharged steam, and to a great extent mixed with steam, ready, immediately any escape is opened from the boiler, to rush out *en masse*, and in proportion to the suddenness of opening, so will be the violence of priming.

The amount, or rather relation of steam space (or steam chest), to the absorbing or using power of cylinders, has, most undoubtedly, a great influence on the priming proclivity of boilers. If the space is inadequate, a vortex will be induced by a species of pumping action, caused by the influx of steam into the cylinders, and the capacity of steam room in the boiler and steam pipes, ought to be sufficient, at least, to preclude any intermittent action on the normal pressure in steam chest, from the alternate closing and opening of the valves, either slide, or expansion, and indeed, if possible, should be sufficient to provide for a slight increase of speed, without suddenly lowering the statical pressure in boilers. In the arrangements of boilers for vessels of war, constructors no doubt will meet with enormous difficulties in finding ample space for all their requirements, but sufficient steam space, certainly is a *sine qua non*, if disaster is to be avoided, and we think this should never be less than 10 or 12 times the steam used in cylinders at each stroke of priming.

The degree of homogeneity of the water in the boilers, will frequently affect or cause priming. Every one knows that nearly all saline solutions require a higher temperature to boil than fresh water, and in water of very high saline density the boiling point itself is exceedingly hard to determine. Regnault mentions some cases of highly concentrated solutions, in which the boiling point of the thermometer proceeded by jumps and starts, so much as to indicate variations on the centigrade thermometer of $10°$. But even under the usual density at which boilers are worked there is some irregularity, and this is further influenced when the homogeneity of the boiler water is affected by feed of different density, and far more so, when the feed contains foreign matter. We have all witnessed, frequently, the circumstance in getting into water of a different character, such as is met with in entering and leaving rivers or their estuaries, that this change produces violent priming: indeed, almost any circumstance that rapidly alters the boiling point, such as the introduction of fresh water as being more volatile, or of water containing foreign matter not in chemical combination, will inevitably lead to the same result. In this latter case, the vapour at the moment of its formation, has only to overcome the attractive force within itself, the same as pure volatile water, and not the adhesions and special affinities of the dissolved saline particles as in sea-water. We mentioned previously, that suddenly opening the communication by reducing the pressure, would lower the boiling point, and induce severe priming; and even when the pressure accumulates in getting up steam or when standing, as it must above the normal load on the safety valve, to overcome its vis inertia and open it automatically, a similar result follows, frequently large volumes of water being sent right up the waste steam pipe on to the deck; and if the engines are suddenly started without regard to the pressure not being equal to the load on the safety valve, the priming comes into the cylinders, if in excess, jeopardizing the engines, and if only moderate, as in the form of surcharged steam, seriously impairing the development of power and speed. How often have we seen engines brought up several revolutions from these causes? It may be asked—What is the practical remedy for all this? We shall, in conclusion, venture on several suggestions, to obviate, or at least mitigate, the worst effects here mentioned: but our immedate object is, more especially, to call attention to, and properly refer effects to their legitimate causes, believing, as we do, that the vigilance and experience of most engineers is competent to deal best with the difficulty when it arises, only insisting that constructors should take account of these serious obstacles with which engineers have so frequently to contend, and by structural and other contrivances, assist them in all their power.

FIG. 4.

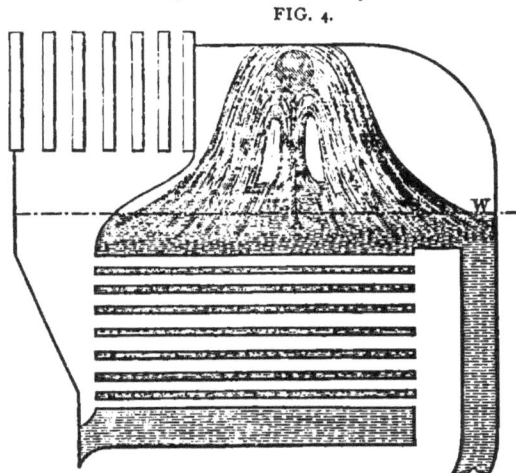

Amongst these constructive arrangements we have already adverted to, the form of the boiler in reference to relative area of surface, and perhaps the next in importance, is the position and extent of internal steam pipe. It is too frequently the practice to limit this in absorbing capacity too much, and to localize it also too much; the internal steam is usually carried along the centre of the steam chest in the manner shown in the accompanying woodcut, (Fig. 4), the egressing steam may be presumed to follow the lines here shown, in which case a vortex, at A, shown by the arrows, is inevitable, carrying water with it from the surface, similar almost in all respects to a water spout.

If two or more internal steam pipes were fitted, all leading to the same outlet, the absorbing area would be increased, and this vortex obviated; in addition to which, perforated zinc intercepting plates, might advantageously be fitted midway betwixt the water surface, and the pipes, which would probably arrest and throw down by its own gravity, the watery particles with the steam ; we prefer zinc plates because they might act as the positive couple in the electric action arising from the brass tubes in the boilers, or the copper tubes in the surface condenser, making the iron a negative or positive, and thus save the destructive action on the boilers. Be that however as it may, there can be little question that it would assist us in getting purer steam, and considerably arrest priming, either of a violent or a chronic form.

As further developing this process of arresting the watery particles, we are strongly of opinion that each boiler should be supplied with a distinct "Separator," fitted betwixt the steam outlet or boiler steam chest, and the superheater ; we know the beneficial action of our general separator, but there is just so much positive loss of heat in allowing the water to traverse thus far, and then blowing it over-board— Why not arrest the hot water at its earliest stage, and let it drain again into the boiler, which it obviously would do by its own gravity ? There is here the further important consideration of the damaging effect induced in the "superheater," when, instead of performing its legitimate function, of raising the temperature of the steam alone, it becomes converted into a kind of supplementary boiler. We take the liberty of strongly urging these considerations on constructors, and the Naval authorities, feeling convinced, if economy is to be realised, we must look to the boilers to supply us with pure steam. There are quite sufficient refrigerating influences acting on the steam in its passage to the engines ; and power, and safety in the machinery is dependent to a very great extent on this quality of the steam. We have *Steam* Engines to supply with food, not Steam and Water Engines, and our success in economy will certainly be in proportion to our compliance with this rigid and paramount condition.

OXYDATION IN BOILERS—CAUSE AND PREVENTION.

OXYDATION IN BOILERS.—CAUSE AND PREVENTION.

THE first introduction of surface condensers was attended by startling effects produced on the boilers, so much so, that the United States Government Steamer, " *Dacotah*," became corroded through nearly one-third of their thickness in three months; and, indeed, the cases are innumerable and existing up to this time, of serious and damaging action going on in boilers from surface condensers. The points of attack are varied to some extent, but seem in all cases to affect most dangerously, hanging surfaces—we mean surfaces with the faces downwards—these faces seem generally to be eaten into small pits, which would of course eventuate in perforations; the parts thus attached are under furnaces, under smoke boxes, and inside top of steam chests; while sides of furnaces, boiler sides, and on other vertical surfaces, the action is more uniform and diffused ; of furnaces and other surfaces exposed to considerable heat, the action appears to exhibit itself in the form of blisters, consisting of an outer-coating of lime, with a lining attachment of oxide of iron, and we have witnessed recently several most novel cases, so far coincident in their appearance, and only with surface condensers, as to justify its being referred to this common cause ; this appearance consisted of very extensive depositions of a spongy calcareous nature, irregular in shape, and about the size of nutmegs closely packed, as if they had been pelted at the back end of furnace top, and lower part of back tube plate, so thick indeed had they formed after a few weeks' steam-ing as to jeopardise the safety of the plates. This latter feature be-comes very important in its consequences, as the Hydrometer does not give any indication of the probability of such deposition. What is the connection betwixt this phenomena and the destructive oxydation as referable to the presumed galvanic action conveyed by the same water coming so frequently in contact with the copper tubes in surface condensers? Perhaps it may be slightly remote, but it is, we think, to be regretted, that opportunities are not afforded of making more conclusive experiments and observations on the circumstances as they occur in practice; all these effects, and indeed everything that is now found to exhibit itself in boilers using feed from surface condensers, is referred without further enquiry or doubt to the general cause of "galvanic action." Oxydation seems palpably due, however to this cause, but whether electrical phenomena, as understood in distinction to galvanic action, may not play a very important part in explanation of the other exhibition referred to, remains to be demonstrated, and without at present, attempting any further physical development of these peculiar depositions, we will try to recapitulate the influences likely to induce galvanic action and concurrent oxydation, and suggest that which appears to us as the obvious and easily attained practical remedy.

For years it has been notorious how rapidly marine boilers are destroyed, and in men-of-war this deterioration is of more vital consequence than the mere cost of removal; the efficiency of the vessel is impaired, and not unfrequently the services of the vessel completely lost from this, and with the introduction of surface condensers, and higher pressure now used, the destructive action appears to be gaining in virulence, and if some means are not adopted for arresting this form of attack, the most valuable sources of economy and power in steam engines, namely, surface condensers and superheating, will be brought into some odium, and their extension and development seriously checked.

It must be remarked, that there is one characteristic attached to the surface condenser, whether fitted with copper or any other tubes, the effect of which in itself we have no adequate or sufficient information, that is, the repeated distillation of the same water, and its being continually in contact with the boiler; is there any molecular modification of its atoms, sufficient to account for the exceptional appearance of the boiler under these conditions is a question that must obviously deserve consideration; or if the water by cohobation attains any special coercive properties sufficient to eject summarily when in contact with highly heated surfaces, the saline ingredients contained in extra feed, and spontaneous additions brought from the cylinders.

The peculiar nature of the depositions above referred to would seem almost to suggest some ground for this hypothesis, as they resemble in appearance a compound of over-heated tallow in steam with probably some of the fibrous elements of the packing, combined with the usual calcareous matter, except that these appearances are only exhibited in connection with surface condensers. This reference to the subject would more properly have been treated under the heading of " Incrustrations," but leaving this highly interesting study, in which so much suggests itself sufficiently important to provoke further investigation, such are the respective properties of furnace boilers, under their high temperature, and the saline and other foreign matter before referred to, as to whether they are paramagnetic, neutral, or dramagnetic toward each other, as it must be remembered that these saline precipitations never occur in cold boilers, although the point of saturation may far exceed that at which the precipitations occur in boilers when heated and evaporating steam. Reverting to the more particular and large subject implied by our heading to this chapter, we referred at starting to what appeared to be the most general and prominent points of attack, and the irregular mode of its action on some of these parts ; this irregularity may possibly be explained on mechanical grounds, that is, the adhesion of watery particles in globules, charged with electro force, when the surrounding parts have become dry, and therefore uninfluenced by an electric medium, but the general body of the boiler is by no means allowed to escape the silent and persistent action ; we have recently seen a plate in the steam chest of a boiler, but three years in use, *wasted* away from ⅜ in thickness to ⅛, and another plate on the sides of the furnaces of another vessel were uniformly wasted to about half its former thickness in less than three years ; these cases might be multiplied, and further experience and investigation will show how important and extensive this action is. We think it would be entirely supererogatory to attempt to refer this to any other cause than a powerful and continuous electric force being excited by the contiguity of large masses of copper and iron, and possibly assisted in its effects by induced thermo-electric currents in the boiler itself.

We remarked on the circumstances that the same water was being continually converted into steam, and re-converted into water, continually in contact with the same *galvanic pile* composed of the copper steam and feed pipes, copper and condenser tubes, here forming the negative pole, and the iron in the boilers, condenser, and hot wells, furnishing the positive element; these necessary heterogenous sub-stances, with a saline solution as a medium, will obviously develope an amount of electro force in proportion to the surfaces in contact, the activity of the liquid medium, or the assistance received to its action from thermal influences in the boiler. There is one feature in the exhibition of positive effects, that is proper to be observed; that is, how very much in excess the positive polarity of the iron in the boilers is to that of the iron in the condensers and hot well. This may be influenced by two causes, the most obviously suggestive, perhaps, being the difference inthe character of the liquid, it being presumably, fresh water, as a medium in the condenser hot well, while it is a saline solution in contact with the boiler. The known thermo-electric forces encited by the junction of the copper tubes in *iron* tube plates, with a strongly marked differentive temperature, would most likely furnish a further cause for this difference in effect in the two localities referred to. We do not forget the electric force, possibly excited by the friction of the water in ebullition, against the sides of the boiler, or the force called into play by the disengagement of the saline constituents in sea-water at the moment of forming steam; these causes exist in all boilers, and their influence is too complex in itself, and so likely to be mitigated by contrary influences from the same cause, that we think, practically, they may be disregarded here, and thus limit the assignable causes: thus, primarily, the contiguity and connection of large copper and iron surfaces, with an exciting medium continually obtaining accessions of specific inductive capacity; and secondly, the thermo-electric force developed in the tube and tube plate. The electro motive force from this cause, in a good-sized boiler, may be calculated according to Dr. Mathiessen's tables, as equal to about two cells of Dancell's battery.

In referring to these various causes of oxydation, we have not mentioned those easily preventible ones, arising from keeping boilers damp. This is a fruitful cause of deterioration common to all boilers and any engineer we should regard as highly culpable, who allowed boilers to remain damp for a moment longer than avoidable: better by far keep them full of water when fires are out, than merely blown or pumped out, without being immediately and thoroughly dried. We should almost think it impertinent drawing attention to so obvious a cause of oxydation, but that our experience has made us aware of so frequent an inattention to this necessary precaution. To neutralize the action of the copper on the iron of the boilers, the interposition of zinc surfaces would, no doubt, prove most effective. This fact has been demonstrated frequently as a purely physical fact, and has been tried very successfully in steam boilers ; indeed, Mr. Hay, the Admiralty chemist, submitted a plan to the Admiralty, ten or fifteen years ago, for the application of zinc under several different modifications, and for which the Admiralty took out a patent But we believe to this day, no ship in the Navy has ever been fitted with any contrivance of the sort. We proposed in a former chapter *(on priming)* the fitting zinc intercepting plates, and we are tolerably sanguine that these plates would prove highly beneficial, as forming a more positive plate than iron, and thus free the boilers from attack, by rendering them electrically passive. It is not necessary in this place to demonstrate the precise mode in which this action takes place, but would refer those of our readers who wish to master the details of "current electricity" to *Ferguson's Electricity*, or *Ganot's Physics*, or many other, no doubt, excellent works on the subject. Perhaps on some future occasion we shall have more information to offer as to the effect of the practical remedy here suggested, at present it would be premature to conclusively and absolutely affirm its efficiency without further time for observing the result of its application.

Apart from this introduction of zinc in contact with the boilers, there can be no question that very much might be effected by tinning all the condenser tubes. Internal steam and feed pipes might easily be tinned both outside and in; or, no doubt, a cheap resinous solution might be obtained, that would withstand the influence of temperature, and at the same time, thoroughly insulate these copper pipes and stop their action on the boilers, and would at the same time be easy of application even to the *inside* of condenser tubes—be cheap, readily applied, and renewed. We think a great mistake is committed in keeping condenser tubes too clean: the slight coating of mixed tallow and worn gland packing, which attached itself to the tubes, in a vessel under our immediate observation, was found most effectually to insulate them, if not to produce a condition of resinous electricity, which is a negative quantity, and save the boilers from any destructive action, and without any sensible effect on their refrigerating or condensing power. In the other cases we mentioned, where the plates had been' so much wasted, we happen to know that more than usual pains have been taken by the application of caustic potash, and the free use of the "blow through" to keep the condenser tubes clean: and further, it is notorious, that boilers that are running hard and continuously, are not affected so much, in proportion, as those boilers steaming occasionally only. Whether this arises from electric polarization of the tubes after the first action, or from partial insulation, as above referred to, it is difficult to say, but in easy working steamships, where the time is ample to keep the tubes clean and free, the temptation to do so is perhaps excusable, considering that in the Royal Navy at least, it is regarded by the authorities as a duty to keep them so. It is, perhaps, doubtful whether it would be desirable to protect the sides and steam chests of boilers with any coating, to protect them from action. If so, probably, a coating of concentrated nitric acid would stand the best, as it would raise an insoluble oxide on the surface; this would ensure its passivity, and perhaps render it electro-negative, and it is equally to be questioned if the "thermo-electric" forces could practically be dealt with. By antimony washes introduced on each boiler tube, they would be difficult to replace, and possibly lead to other complications not advantageous. However, by referring to this matter, we keep within the limit of our intention throughout these remarks, that is, to render them more suggestive of a line of investigation to be followed, than an exhaustive treatment of the subject.

We may mention in conclusion, that we have seen tried, a galvanic feed box, the patent of Mr. Footman, which is a step in the right direction; but it seems to us to be too deficient in surface power to be of much use in effecting its object; however, its mechanical arrangement admits of a great extension of surface, and if the patentee avails himself of this facility, it may prove highly beneficial.

We hope we have been successful in pointing out the cause of a serious drawback to the development of, perhaps, one of the most useful and economical auxiliaries in steam engines, and equally not failed in indicating the remedy: and we venture to think that indifference on the part of those professional engineers who have any responsibility or power to correct the evils, or, at least to inaugurate tangible and practical investigation would be highly culpable, not only in the interests of scientific progress but in saving the pockets of proprietors. As Mr. Armstrong observes, "engineers are too apt to let these matters get into the hands of chemists and druggists, instead of applying their own talent and experience to overcome or mitigate the sources of loss and waste so apparent in practice."

PRACTICAL SUMMARY ON THE MANAGEMENT OF BOILERS & SUPERHEATERS.

PERHAPS the paramount duty of an Engineer on watch is to ensure the water being kept to the proper height in the boilers, both as to absolute height and uniformity in maintaining this normal level; and as nearly all vessels are fitted with brine cocks, or surface blow off, this becomes a matter of easy adjustment, as more water should be kept in the glass when the ship has any considerable motion than when still, about one-third up the glass will be quite ample, when the ship is uniform in her trim; if the water surface rises too high it is apt to be carried along into the engines as surcharged steam, from mere contiguity, and also because the greater the height the escaping steam bubbles have to rise through the water, the more they increase in volume, and thus carry larger adhesions of watery particles : these reasons are apart from the consideration of reducing the steam space before referred to, and is no fancy cause of mild priming, but a well established and constant cause. Another reason for keeping the water as low as practicable is the better command of steam it gives the engines.

The *density* is another paramount matter requiring constant attention, and of course the brine and feed cocks being properly adjusted by very slight alterations when required, both the normal level of the water and its proper density may be kept uniform without violent or considerable changes ; indeed, the Engineer ought always to avoid making great additions in the amount of feed, as the steam, even with the best managed fires, cannot be kept uniform, and the stokers are apt to get discouraged and indifferent when they observe that their efforts are not efficiently seconded. In boilers using fresh water feed from surface condensors, about 15° by Hydrometer will be found the best density, and for boilers using the sea-water from ordinary condensors, 22° by Hydrometer is the best mean, observing that it should never rise to 25° ; it is a great mistake as we before observed to keep the point of saturation too low.

We believe sufficient attention is not paid in working with large grate surfaces—we do not mean here absolutely large, but large in proportion to the steam required to be produced in keeping the bars thoroughly covered ; it is completely throwing coals away, allowing space on the bars for cold air to pass immediately through into the tubes, without going through the fire, there need be no apprehension of getting steam too high, or too much of it with the bars well covered. The ash pit doors will always efficiently control the combustion, and if it is necessary to check the production of steam when the engines are stopped suddenly, we see no objection to opening the smoke box doors, a means almost as efficient for safety as drawing the fires, with the advantage of immediate promptness.   When water suddenly leaves the boiler in violent priming, combustion is hereby immediately arrested, and the tubes cooled by the rush of air through them.

The number of boilers used on any occasion ought obviously to be dependent on the consumption of coal, which " cæteris paribus " is relative to the amount of steam required.   Thus, usually in practice, the consumption of coal is the first rigid circumstance to take into consideration in determining the number of boilers required; and in deciding this question, we have only to regard the amount of grate surface best adapted for using this fixed amount of coal most economically ; we are aware that on this point there is a great difference of opinion amongst practical Engineers, we have so frequently met with men of undoubted experience advocating almost an indiscriminate large surface to be used, with a view of absorbing as they call it, all the fuel—indeed we know of nothing which may be said to connect itself with the daily practice and experience of an Engineer, on which wrong views are so general or so persistently held. Our object is not to write a treatise on boilers here, but we commend strongly *Armstrong's Rudimentary Treatise on Steam Boilers*, Chapter IV, to those Engineers who insist on reversing the conditions required in a tubular boiler, in trying to assimilate them to a Cornish boiler.   We have tried the experiment repeatedly under practical circumstances, and demonstrated enough to convince the most sceptical that there is no economy, but loss, in reducing the consumption of coal *below 10 lb. per square foot of grate surface,* and if superheating is to be effective, the rate per square foot of grate may be advantageously increased to 13 lbs., and in all cases the rate of consumption per square foot ought to be the sole consideration in determining the number of boilers to be used. Perhaps it is a misfortune that no *ready* and efficient means are provided for temporarily shortening furnaces ; about one-third, "bricking up" is almost too permanent a measure if the full resources of the boiler are required promptly and urgently. Fires should always be moderately brisk, always level, not too thick, and never permitted by any possible chance to burn into holes. If fires are kept tolerably level, careful attention to the ash pit doors, with uniformity in feed and brine, will enable the Engineer to maintain his steam guage at its normal or ordered height. If all the boilers are not in use, care should be taken to see the furnace and ash pit doors of the annexed boilers tightly closed.

Superheaters should be kept drained, and as the Admiralty have fixed the maximum temperature of superheated steam at 300°, we have every confidence that the limit is fixed on sufficient data to pro- vide all the economy to be realised from superheating, with ·a due regard to saving the working surfaces. We have a strong objection to "thumb rules," but from observing the limit at which we found the attrition getting threatening as to its results, we think this even a little too high.

To guard against priming under ordinary circumstances we have said enough already, and also, as to the measures to be adopted when changing into rivers. Feed should be used sparingly, and large accessions which induce violent action avoided.

It may possibly appear to many exceedingly elementary in charac- ter and unnecessary, in fact, to give details of management on such minute and obvious points as are here briefly referred to, but our own observation—and we have no doubt the experience of many responsible senior and chief Engineers—furnish innumerable instances of inatten- tion and neglect of these every day requirements, that we are apt to wonder whether "zeal and attention" or the knowledge of duty, is the deficient feature; however, the fact of these matters being so frequently not attended to, and their importance, must be our apology for going beyond our intention in giving advice thereon.

58

TABLE of Specific Gravities of Brines of Sea-Water, with corresponding Weights of Constituents and Densities, &c. [*Table* A 1.

1	2	3	4	5	6	7	8
Specific Gravity at 60 Fahrenheit.	Weight of Salt in a cubic foot of compound	Weight of pure water in a cubic foot	Density or proportion of salt sea water 1000.	Per centage of salt.	Boiling Temperature.	Corresponding indication on Naval Hydrometer 900	Capacity for heat, equal weights.
1.0138	17.3	996.5	.466	1.7	212.6	5.	.986
1.0156	20.0	995.6	.539	1.97	212.76	5.5	
1.017	21.9	995.1	.590	2.15	212.83	6.1	
1.018	23.3	994.7	.628	2.29	212.88	6.5	.981
1.019	24.6	994.4	.662	2.414	212.93	9.6	
1.020	26.1	993.9	.704	2.56	212.99	7.2	.979
1.021	21.5	993.5	.741	2.69	213.04	7.6	
1.022	28.9	993.1	.778	2.74	213.06	7.9	.978
1.023	30.3	992.7	.816	2.96	213.14	8.3	
1.024	31.8	992.2	.850	3.10	213.2	8.7	.975
1.025	33.2	991.8	.892	3.24	213.25	9.0	
1.026	34.6	991.4	.932	3.37	213.3	9.4	.973
1.027	36.1	990.9	.970	3.515	213.36	9.7	
1.0277	37.12	990.58	1.000	3.61	213.4	10.0	.971
1.028	37.55	990.45	1.011	3.65	213.41	10.1	.970
1.029	39.00	990.00	1.05	3.79		10.5	
1.030	40.45	989.55	1.09	3.93	213.52	10.8	.968
1.031	41.90	989.10	1.129	4.06		11.7	
1.032	43.35	988.65	1.168	4.20	213.63	11.5	.966
1.033	44.80	988.20	1.207	4.34		11.9	
1.034	46.26	987.74	1.246	4.47	213.73	12.3	.964
1.035	47.76	987.24	1.285	4.61		12.6	
1.036	49.18	986.82	1.325	4.75	213.84	13.0	.962
1.037	50.64	986.36	1.364	4.88		13.3	
1.038	52.10	985.90	1.403	5.02		13.7	.960
1.039	53.56	985.44	1.443	5.15	214.00	14.1	
1.040	55.02	984.98	1.492	5.29		14.4	.957
1.041	56.48	984.52	1.521	5.425	214.1	14.8	
1.042	57.94	984.06	1.560	5.56		15.2	.955
1.043	59.40	983.60	1.587	5.695	214.2	15.5	
1.044	60.88	983.12	1.613	5.83		15.9	.953
1.045	62.36	982.64	1.666	5.97	214.31	16.2	
1.046	63.84	982.16	1.720	6.10		16.6	.951
1.047	65.34	981.66	1.760	6.24	214.4	17.0	
1.048	66.58	981.15	1.800	6.38		17.3	.949
1.049	68.36	990.64	1.841	6.52	214.5	17.7	
1.050	69.87	980.13	1.822	6.65		18.0	.947
1.051	71.38	979.62	1.921	6.79	214.63	18.4	
1.052	72.89	979.11	1.960	6.63		18.8	.944
1.0528	74.24	978.56	2.000	7.05	214.7	19.0	.943
1.053	74.40	978.60	2.004	7.065		19.1	
1.054	75.90	978.10	2.045	7.201	214.79	19.5	.942
1.055	77.40	977.60	2.085	7.336		19.8	
1.0554	77.92	977.48	2.099	7.390	214.86	20.0	.941
1.056	78.91	977.19	2.126	7.472		20.2	

1	2	3	4	5	6	7	8
Specific Gravity at 60° Fahrenheit	Weight of salt in a cubic foot of compound.	Weight of pure water in a cubic foot.	Density or proportion of salt, sea water 1.000.	Per centag of salt.	Boiling Temperature.	Corresponding indication on Naval Hydrometer, 90°.	Capacity for heat, equal weights.
1.066	94.08	971.92	2.559	8.823	215.42	23.8	.929
1.067	95.60	971.40	2.587	8.966		24.2	
1.068	97.12	970.88	2.616	9.108	215.53	24.5	.927
1.069	98.65	970.35	2.657	9.235		24.9	
1.070	100.17	969.83	2.698	9.362	215.62	25.3	.925
1.071	101.70	969.30	2.739	9.493		25.6	
1.072	103.23	968.77	2.780	9.630	215.73	26.0	.923
1.073	104.76	968.24	2.822	9.764		26.3	
1.074	106.3	967.70	2.864	9.897	215,83	26.7	.921
1.075	107.83	967.17	2.904	10.030	215.9	27.0	
1.076	109.36	966.64	2.946	10.163	215.9	27.4	.919
1.077	110.90	966.10	2.988	10.297		27.8	
1.0773	111.36	965.94	3.000	10.337	216.0	27.9	.917
1.078	112.42	965.58	3.028	10.425		28.2	
1.079	113.95	965.05	3.076	10.560	216.09	28.5	.915
1.080	115.48	964.52	3.124	10.693		28.9	
1.081	117.00	964.0	3.158	10.824	216.2	29.3	.913
1.082	118.53	963.47	3.193	10.955		29.6	
1.083	120.06	962.94	3.234	11.086	216.3	30.0	.911
1.084	121.60	962.40	3.276	11.218		30.3	
1.085	123.13	961.87	3.317	11.348	216.4	30.7	.909
1.086	124.60	961.34	3.358	11.479		31.1	
1.087	125.10	960.90	3.398	11.605	216.5	31.4	.907
1.088	127.63	960.37	3.438	11.731		31.8	
1.089	129.16	959.84	3.479	11.860	216.6	32.2	.905
1.090	130.70	959.30	3.521	11.990		32.5	
1.091	132.23	958.77	3.562	12.12	216.7	32.9	.903
1.092	133.77	958.23	3.603	12.250		33.2	
1.093	135.31	957.69	3.644	12.380	. 216.8	33.6	.901
1.094	136.88	957.14	3.685	12.510		34.0	
1.095	138.40	956.60	3.727	12.639	216.9	34.3	.899
1.096	139.94	956.06	3.770	12.768		34.7	
1.097	141.48	955.52	3.811	12.897	217.0	35.0	.897
1.098	143.03	954.97	3.853	13.026	217.0	35.4	.896
1.099	144.58	954.42	3.895	13.155		35.8	
1.100	146.14	953.86	3.937	13.285	217.1	36.1	.894
1.101	147.70	953.30	3.978	13.415		36.5	
1.1015	148.480	953.02	4.000	13.480	217.2	36.7	.892
1.102	149.2	952.8	4.02	13.54		36.8	
1.103	150.7	952.3	4.06	13.66	217.3	37.2	.891
1.104	152.3	951.7	4.10	13.79		37.6	
1.105	153.9	951.1	4.15	13.92	217.4	38.0	.888
1.106	155.5	950.5	4.20	14.05		38.3	
1.107	157.1	949.9	4.24	14.19	217.5	38.7	.886
1.108	158.7	949.3	4.29	14.32		39.0	
1.109	160.3	948.7	4.33	14.45	217.6	39.4	.884
1.110	161.90	948.1	4.37	14.58		39.8	
1.1107	163.00	947.7	4.39	14.67	217.7	40.0	.882
1.120	177.8	942.2	4.79	15.87	218.1	43.3	.873
1.125	185.0	939.4	5.00	16.50	218.4	45.1	.868
1.385	207.0	931.5	5.57	18.17	219.1	50.0	.854
1.150	225.5	924.5	6.07	19.60	219.7		.843
1.160	241.5	918.5	6.49	20.77	220.1		.834
1.170	257.0	913.0	6.92	21.96	220.6		.824
1.180	273.0	907.0	7.35	20.313	221.0		.816
1.190	289.0	901.0	7.78	24.28	221.5		.806
1.200	305.0	895.0	8.21	25.41	222.0		.796

	1·1400	1·13500	1·1300	1·1250	1·1200	1.1150
60	1·1400	1·13500	1·1300	1·1250	1.1200	1.1150
70	1·1382	1·1332	1·1282	1·1233	1.1183	1.1133
80	1·1363	1·1313	1·1263	1.1214	1.1164	1.1114
90	1·1340	1·1291	1·1242	1·1192	1.1143	1.1093
100	1·1314	1·1265	1.1215	1·1166	1.1117	1.1068
110	1·1288	1.1239	1·1190	1·141	1.1092	1.1043
120	1·1257	1·1208	1·1160	1·111	1.1063	1.1014
130	1·1228	1·1179	1·1130	1·1082	1.1033	1.0970
140	1·1194	1·1146	1·1098	1·1050	1.1002	1.0953
150	1·1158	1·111	1.1062	1·015	1.0967	1.0919
160	1·1122	1·1074	1·027	1·0979	1.0932	1.0894
170	1·1084	1·1037	1·0990	1·0942	1.0895	1.0847
180	1·1045	1·00998	1·0951	1·0903	1.0857	1.0810
190	1·1005	1·0958	1·0912	1·0864	1.0818	1.0770
200	1·0965	1·0919	1·0872	1·0825	1.0779	1.0732

DILATION OF SEA-WATER BRINES BY HEAT,

AT TEMPERATURE FROM 60° TO 200°.

SPECIFIC GRAVITIES.

Temperature.											
60	1.0850	1.0800	1.0750	1.0700	1.0650	1.0600	1.0550	1.0500	1.0450	1.0400	1.0350
70	1.0835	1.0785	1.0735	1.0685	1.0635	1.0586	1.0536	1.0486	1.0436	1.0386	1.0337
80	1.0817	1.0768	1.0718	1.0669	1.0619	1.0570	1.0522	1.0471	1.0421	1.0372	1.0323
90	1.0797	1.0748	1.0699	1.0650	1.0600	1.0551	1.0502	1.0452	1.0403	1.0354	1.0305
100	1.0774	1.0725	1.0675	1.0626	1.0573	1.0529	1.0480	1.0431	1.0382	1.0333	1.0284
110	1.0750	1.0703	1.0653	1.0603	1.0556	1.0507	1.0458	1.0409	1.0360	1.0311	1.0262
120	1.0723	1.0674	1.0625	1.0577	1.0524	1.0480	1.0432	1.0383	1.0334	1.0286	1.0237
130	1.0695	1.0646	1.0608	1.0549	1.0501	1.0453	1.0405	1.0357	1.0308	1.0260	1.0211
140	1.0664	1.0616	1.0568	1.0520	1.0472	1.0424	1.0376	1.0328	1.0280	1.0232	1.0184
150	1.0632	1.0584	1.0536	1.0488	1.0440	1.0392	1.0344	1.0296	1.0249	1.0201	1.0153
160	1.0598	1.0550	1.0503	1.0455	1.0407	1.0359	1.0312	1.0264	1.0217	1.0169	1.0121
170	.1.0563	1.0515	1.0468	1.0420	1.0372	1.0325	1.0278	1.0231	1.0183	1.0136	1.0088
180	1.0621	1.0478	1.0431	1.0384	1.0337	1.0290	1.0243	1.0196	1.0149	1.0102	1.0054
190	1.0488	1.0441	1.0394	1.0347	1.0301	1.0254	1.0207	1.0160	1.0114	1.0067	1.0020
200	1.0450	1.0404	1.0357	1.0310	1.0264	1.0217	1.0170	1.0123	1.0072	1.0030	0.9984

DILATION OF SEA-WATER BRINES BY HEAT,

AT TEMPERATURE FROM 60° TO 200°.

SPECIFIC GRAVITIES.

Temperature								
60	1.0300	1.0270	1.0250	1.0200	1.0150	1.0100	1.0050	1.000
70	1.0287	1.0264	1.0237	1.0187	1.0137	1.0088	1.0038	.9980
80	1.0273	1.0250	1.0223	1.0173	1.0123	1.0074	1.0025	.9976
90	1.0256	1.0233	1.0206	1.0156	1.0106	1.0058	1.0009	.9960
100	1.0234	1.0211	1.0184	1.0135	1.0085	1.0037	.9990	.9940
110	1.0214	1.0192	1.0166	1.0117	1.0067	1.0019	.9970	.9921
120	1.0118	1.0166	1.0140	1.0091	1.0042	.9994	.9946	.9897
130	1.0163	1.0141	1.0115	1.0036	1.0017	.9981	.9927	.9873
140	1.0136	1.0114	1.0088	1.0040	.9991	.9943	.9895	.9847
150	1.0105	1.0083	1.0058	1.0010	.9961	.9914	.9866	.9818
160	1.0074	1.0052	1.0027	.9980	.9931	.9894	.9836	.9788
170	1.0041	1.0020	.9995	.9948	.9900	.9853	.9805	.9757
180	1.0007	.9986	.9961	.9914	.9866	.9819	.9772	.9724
190	.9973	.9952	.9927	.9880	.9832	.9781	.9736	.9691
200	.9937	.9916	.9891	.9844	.9796	.9751	.9703	.9657

CORRESPONDING TEMPERATURE IN USING NAVAL HYDROMETER. [Table C 1.

Temperature.	10	11	12	13	14	15	16	17	18	19	20	21	22	23	24	25	26	27	28	29	30
60	10.0	11.0	12.0	13.0	14.0	15.0	16.0	17.0	18.0	19.0	20.0	21.0	22.0	23.0	24.0	25.0	26.0	27.0	28.0	29.0	30.0
70	9.5	10.5	11.5	12.5	13.5	14.5	15.5	16.5	17.5	18.5	19.5	20.5	21.5	22.5	23.5	24.4	25.4	26.4	27.4	28.4	29.4
80	9.0	10.0	11.0	12.0	12.9	13.9	14.9	15.9	16.9	17.9	18.9	19.9	20.9	21.9	22.9	23.9	24.9	25.9	26.9	27.9	28.9
90	8.4	9.4	10.4	11.4	12.3	13.3	14.3	15.2	16.2	17.2	18.1	19.1	20.1	21.1	22.1	23.1	24.1	25.1	26.0	27.0	28.0
100	7.6	8.6	9.6	10.6	11.6	12.6	13.6	14.5	15.5	16.5	17.4	18.4	19.4	20.4	21.4	22.3	23.3	24.3	25.2	26.2	27.2
110	6.9	7.9	8.9	9.8	10.8	11.8	12.8	13.7	14.7	15.7	16.6	17.6	18.6	19.6	20.6	21.5	22.4	23.4	24.3	25.3	26.3
120	6.0	7.0	8.0	8.9	9.9	10.9	11.8	12.8	13.7	14.7	15.6	16.6	17.6	18.6	19.5	20.5	21.4	22.4	23.3	24.3	25.3
130	5.1	6.1	7.1	8.0	9.0	10.0	10.9	11.9	12.8	13.8	14.7	15.7	16.7	17.6	18.6	19.6	20.5	21.5	22.4	23.4	24.4
140	4.1	5.1	6.1	7.0	8.0	9.0	9.9	10.9	11.8	12.8	13.7	14.7	15.6	16.6	17.6	18.5	19.5	20.4	21.4	22.3	23.3
150	3.0	4.0	5.0	5.9	6.9	7.9	8.8	9.8	10.7	11.7	12.6	13.6	14.5	15.5	16.4	17.4	18.4	19.3	20.3	21.2	22.2
160	1.9	2.9	3.9	4.7	5.7	6.7	7.6	8.6	9.5	10.5	11.4	12.4	13.3	14.2	15.2	16.1	17.1	18.0	19.0	19.9	20.9
170	.7	1.7	2.6	3.5	4.5	5.5	6.4	7.4	8.3	9.3	10.2	11.2	12.1	13.1	14.0	14.9	15.9	16.8	17.8	18.7	19.7
180	...	.4	1.4	2.3	3.2	4.2	5.1	6.1	7.0	8.0	8.9	9.8	10.8	11.7	12.6	13.6	14.5	15.5	16.4	17.4	18.3
190	...	...	.1	1.1	2.0	3.0	3.9	4.9	5.8	6.8	7.7	8.7	9.6	10.5	11.4	12.3	13.8	14.2	15.2	16.1	17.0
200	...	...	...	...	.7	1.6	2.4	3.4	4.4	5.3	6.3	7.2	8.2	9.1	10.1	10.9	11.9	12.8	13.8	14.7	15.0

SAUNDERS' SALINOMETER.

A simple, efficient and durable self-acting Salinometer or Gauge for showing the density of the water in Marine Boilers, or in Land Boilers where the water used is either salt, brackish, or contains other impurities—such as Lime, Silica, &c., &c., has long been a great want.

The advantages of such an instrument are manifold, as proper attention to the density of the water in Boilers when at work, tends greatly to their preservation and safety, and economy in working, as well as the saving of fuel and expenses of repairs, all of which points are most important to both the users and owners of Steam Power.

Saunders' Patent Salinometer has been designed from practical experience of the failings of other Instruments invented for the same purpose, and much care and

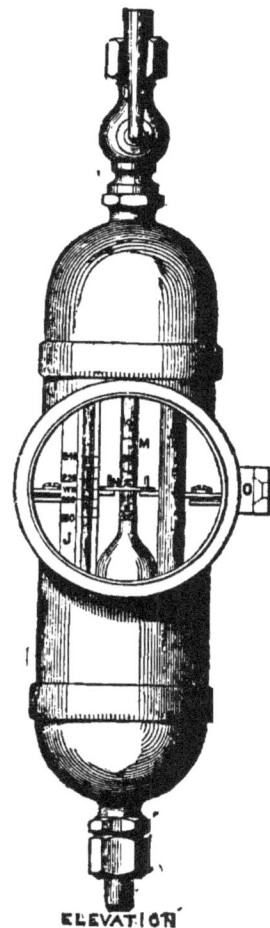

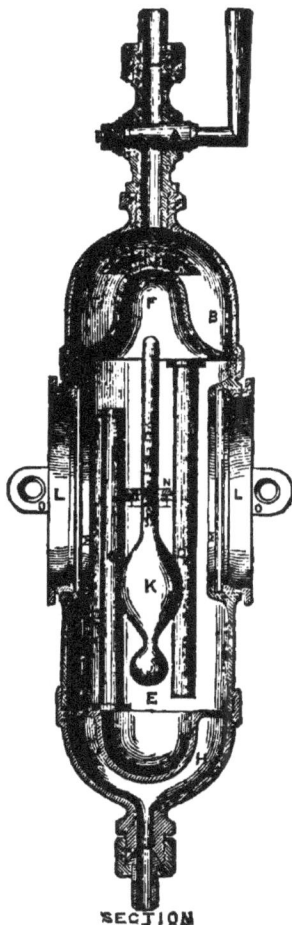

ELEVATION

SECTION

expense have been incurred in perfecting it so as to attain the object sought, viz. :—a thoroughly efficient yet simple instrument, at a moderate cost.

The short description of the Instrument on the other side, with the aid of the above Woodcuts, will explain the principle on which it has been designed, and the advantages it has over every Salinometer yet invented ; and from its simplicity of form and action, any one may readily understand its distinctive features, and be satisfied of its practical utility, and of its meeting the want that has so long been felt.

DESCRIPTION OF SAUNDERS' PATENT SALINOMETER.

THE water from the Boiler is admitted through the straightway Cock *A*, which is the only part of the Instrument that requires any adjustment, and must be set so as to allow a small stream of water to be continuously passing from the Boiler through it, without creating any disturbance of the water in the centre chamber *E*, where the Hydrometer *K* floats. The pipe from this Cock should be connected with the Boiler at about, or a little above, the height of the crown of the furnaces.

The perfect separation of the steam from the water, (which has not been attained by any Instrument yet invented, and without which no Salinometer can indicate truly,) is effected in this Instrument in the separating chamber *B*, by the water striking against the disc *C*, which breaks and disperses it over the chamber and effectually separates it from the particles of steam it may contain.

The water then passes through the inlet pipes *D* into the centre chamber *E*, and all ebullition is entirely prevented by means of the holes in the dome *F* which connect it with the separating chamber *B*.

The over-flow water passes down the outlet pipes *G* through the bottom or discharge chamber *H* into the Bilge, Drain, or other convenient place.

The Thermometer *J* is attached by means of spring clips to one of the inlet pipes *D*, and it and the Hydrometer *K* are completely protected from harm, and can be readily taken out, for cleaning or otherwise, and replaced by simply unscrewing by hand one of the glands *L* and removing the glass face *M*, and this can be effected in a minute as the joints are made with india-rubber rings on each side of the glass faces *M*.

The Hydrometer *K* is read from the top of the bar or guide *N*.

This Salinometer can be fixed either on the Boiler itself or in any convenient place near, by means of the bracket *O*, and being completely enclosed no one can possibly be scalded by it ; and it is entirely self-acting, and after once adjusting the straightway Cock *A*, it will work for months without further attention, and giving as it does a constant index of the state of the water in the Boiler, the quantity and proper time to blow off is arrived at to a nicety, and there is no inducement to neglect or omit the regular examination of, and giving proper attention to the density of the water in the Boiler, which is too often the case, as well as being by the ordinary method a great interference in other duties, and at the same time an unpleasant and tiresome operation.

The Hydrometer in this Instrument is very sensitive and can be thoroughly depended upon, as it floats in still water, and is adjusted at the usual temperature of 200°.

The great importance of the use of a Salinometer cannot be over-stated or too much considered by all who are connected in any way with Steam Power, as it enhances the safety, endurance, and economy in the working of Boilers, and Saunders' Patent is now confidently brought before the public, relying upon the sound principle upon which it is constructed, to obtain for it a general demand.

PRICE.

	£.	s.	d.		£.	s.	d.
Instrument complete with Hydrometer and Thermometer. (See Woodcuts.)	8	8	0	*Thermometers :—*			
				Fahrenheit or Centigrade, each	0	7	6
Do. Do. *Same principle, but of a different pattern*	5	5	0	Glass Hydrometers .. each	0	6	0
				Glass Faces .. .. each	0	1	0
Spare Articles if required.				I. R. Packing Rings for do. per doz.	0	1	0

Wholesale Agent : JAMES HOSKEN, Engineer, 58, FENCHURCH ST., LONDON. E.C.

- From whom they can be supplied.

TROTMAN'S PATENT GALVANIC FEED BOX,

For Boilers of Marine Engines having Surface Condensers.

IT is a well-known fact that, in practice, it is not possible to take advantage to the full extent of the admirable system of surface condensation in marine engines, owing to the injury caused to the Boilers by the galvanic action produced by the copper tubes of the Condenser. From this cause, frequent repairs take place, and vessels are often detained at sea.

In surface condensing engines on long voyages, the water being used over and over again, produces a strong galvanic action on the plates of the boilers, which quickly destroys them. To prevent this, Mr. Trotman has patented what he terms a MARINE FEED BOX, which is so constructed that this galvanic action takes place exclusively within this Box. This desirable result is attained by putting into operation the well-known law of chemical action, arising from the combination of metals of different degrees of galvanic polarity, through the medium of salt water.

Thus, when copper and iron are in galvanic connection, the chemical corrosion is confined to the iron. When iron and zinc are similarly combined, the corrosion is limited to the zinc ; as also when copper, iron, and zinc are in galvanic contact, the zinc only is destroyed.

This invention has been working for nearly five years in the Union Company's Royal Mail Steam Ship " Roman," with great success on long steaming voyages between Southampton and the Cape of Good Hope.

In construction, this Feed Box is extremely simple ; in use it gives no trouble to the Engineer ; and from its efficacy in preserving the Boiler plates from injury, will relieve him from great anxiety in working the boilers ; nor is it necessary when the Patent Feed Box is used, to inject large quantities of salt water into the boiler, so as to form a scale on the plates, a method which is now indispensable to prevent their destruction.

The owners of steam vessels will find that the use of the Patent Marine Feed Box is attended with highly economical results, the necessity for repairs being less frequent, and a great saving of fuel.

With these Feed Boxes in the boilers, the S. S. " Roman," of 1,027 tons, fitted with a pair of surface condensing engines of 220 nominal horse power, steamed from the Cape, (calling at St. Helena and Ascension) in twenty-nine days twenty hours, with an average of twelve tons of coal per day.

No other instructions are requisite for its use than to keep the density of the water in the boiler from 17 to 18 at 120° Fahrenheit, and an examination of the box at the end of each voyage. The efficacy of the Feed Box depends upon the chemical destruction of certain of its parts, and the periodical renewal of these parts is all that is neeessary to ensure its action.

PRICE £15.

Any other information can be obtained by applying to the Patentee's Sole Agent, JOHN DICKENSON, Southsea, Portsmouth, and Bulls' Head Chambers, Manchester.

FCAP. 8vo., PRICE 6s.

RECENT IMPROVEMENTS IN

THE STEAM ENGINE;

Being a Supplement to the "Catechism of the Steam Engine."

BY JOHN BOURNE, C. E.

THE MARINE STEAM ENGINE,

By the Rev. T. J. Main, M.A., and Mr. Thomas Brown, C.E., R.N.

PRICE 12s. 6d.

THE OFFICE AND CABIN COMPANION,

BY J. SIMON HOLLAND.

PRICE 2s.

SEAMANSHIP.

BY COMMANDER G. S. NARES, R. N. Fourth Edition, 8vo., 330 Illustrations, Price 21s.

This Edition is revised and enlarged, and contains four pages of Coloured Flags—comprising Signal Flags, (New Code,) Pendants, Numeral and Alphabetical Flags, Signal Flags used by Men of War for communicating by the Commercial Code, Merchant Signal Flags, Sail Signals, and Beacon Signals.

PORTSMOUTH :—GRIFFIN & CO., THE HARD.

www.ingramcontent.com/pod-product-compliance
Lightning Source LLC
Chambersburg PA
CBHW022001190326
41519CB00010B/1357